INVENTIONS

Build, Build, Build

By Penny Norman, Ph.D

Artwork by Art Huff

Note to Parents

This book is recommended for children eight years and older.

Batteries: The light bulb, telegraph and motor in this kit will work with a 1.5 volt battery. This means that any AA through D-cell alkaline battery will work; they are all 1.5 volts. A D-CELL battery is what we recommend because it lasts the longest. The battery is not included.

Most Common Problems: The most common problem young people encounter is the making of good metal-to-metal electrical contact.

Incomplete sanding of the leads is the most frequent source of problems with the motor. Follow the instructions carefully.

To make the generator work takes some coordination, but it works reliably with practice.

The biggest challenge with the radio is finding a good ground. The metal case ground on the back of a computer works well. For the case to be grounded the computer must be plugged into the wall. With a good ground, the antenna tends to be less of an issue. A simple stretched out length (10 feet (3 meters or more) of insulated wire works well. Students can be their own antenna if they live within range of a clear channel station with a 50,000 watt transmitter. The antenna must be electrically isolated from the ground.

The electromagnet and the telegraph will consume a battery very quickly. Avoid leaving it connected for too long.

SAFETY RULES

This book contains small parts intended for use by children eight years of age or older. The materials may present a hazard if used improperly. There is a choking hazard with small parts. Puncture wounds are possible if the materials are misused. There is a flashlight sized glass light bulb that parents should be sure is treated with care to avoid breakage. The spinning motor and all moving objects must be kept away from the eyes. Read the battery labels for proper usage. The electronic components should not be taken apart. Parental supervision is recommended for safety.

CONTENTS

Go to www.ScienceWiz.com to earn your AWARDS!

Coils, Coils, Coils, wrap a wire around practically anything and you've got a -- COIL!

COIL WRAPPIN
FA DRY
MR.
MARKS OI

Wrap a wire around a nail, you've got a coil.

Wrap a wire around a toilet paper roll, you've got a coil.

Wrap a wire around a marker, you've got a coil.

Slide the coil off the marker, you've got a coil.

Squish the wire together, you've still got a coil.

With a coil you can make a spinning motor, a clicking telegraph, a light flashing generator and a working radio.

Let's get started!

BUILDING A MOTOR

With a coil you can make a working motor. You will need:

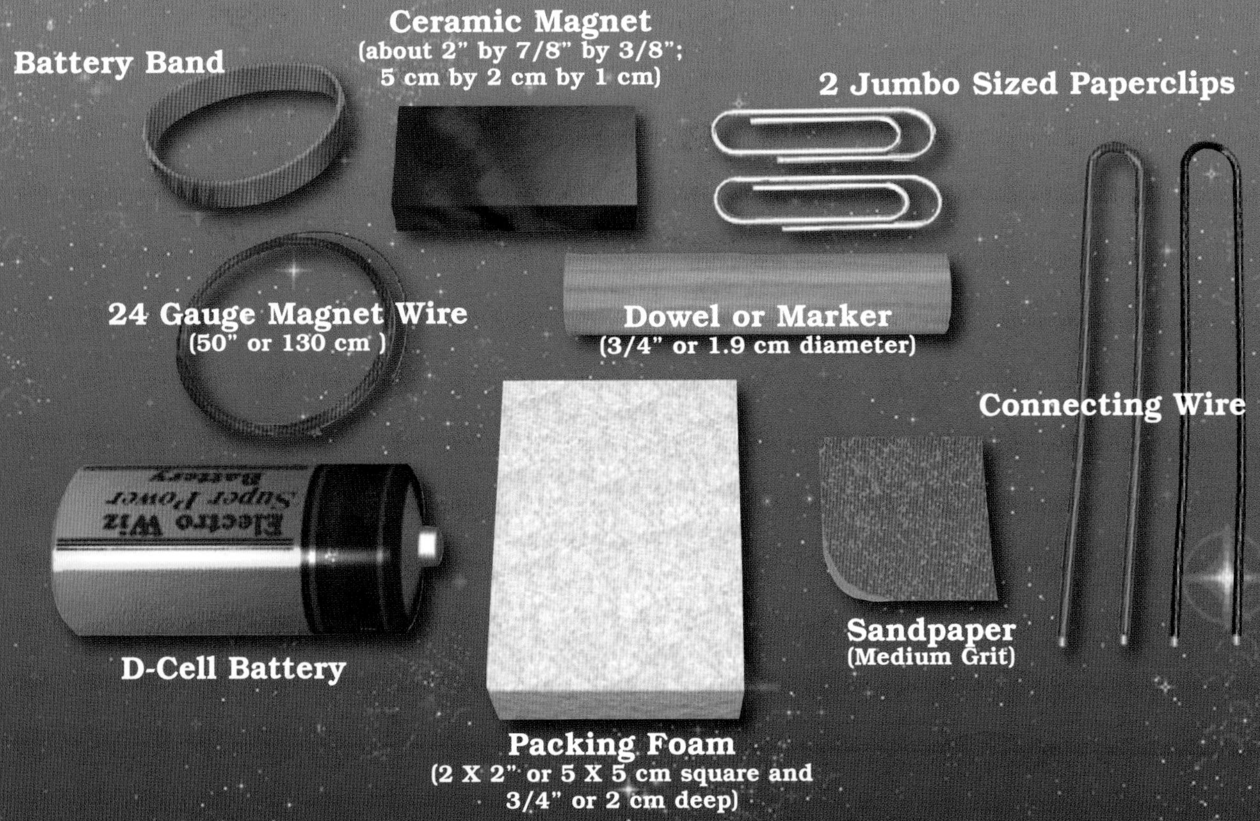
Battery Band
Ceramic Magnet
(about 2" by 7/8" by 3/8"; 5 cm by 2 cm by 1 cm)
2 Jumbo Sized Paperclips
24 Gauge Magnet Wire
(50" or 130 cm)
Dowel or Marker
(3/4" or 1.9 cm diameter)
Connecting Wire
Electro Wiz
Super Power
Battery
D-Cell Battery
Sandpaper
(Medium Grit)
Packing Foam
(2 X 2" or 5 X 5 cm square and 3/4" or 2 cm deep)

You will be winding the magnet wire into a coil.

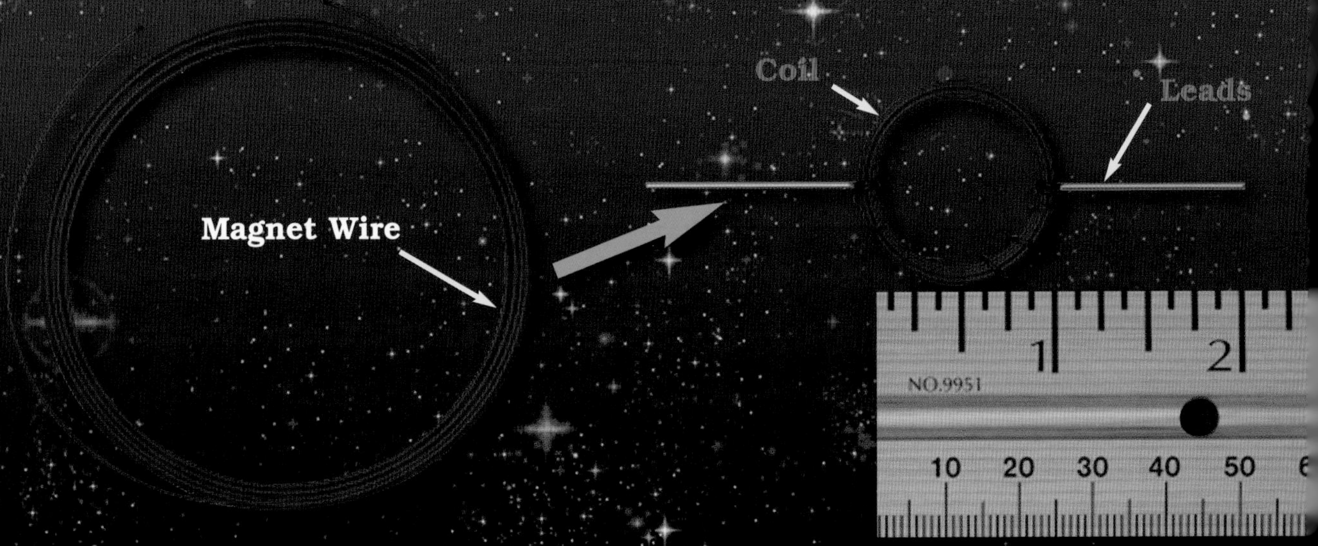
Coil
Leads
Magnet Wire
1
2
NO.9951
10
20
30
40
50

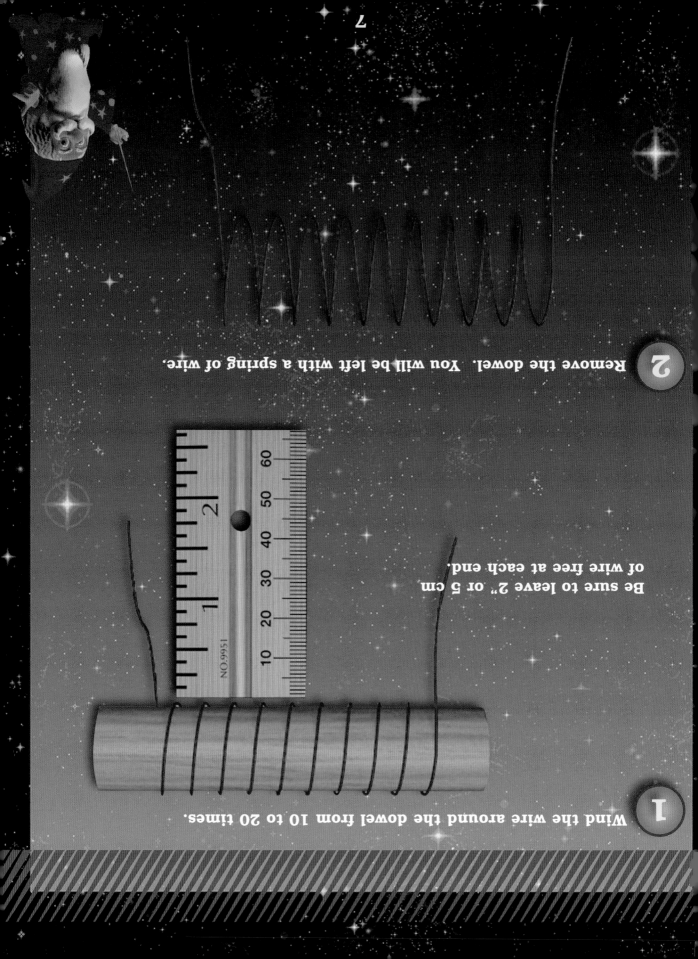

1 Wind the wire around the dowel from 10 to 20 times.

NO.9951
10
20
30
40
50
60
1
2

Be sure to leave 2" or 5 cm of wire free at each end.

2 Remove the dowel. You will be left with a spring of wire.

3 Squeeze the coil together and wrap one end of the wire around the coil.

4 Wind the free end of the wire around the coil like this.

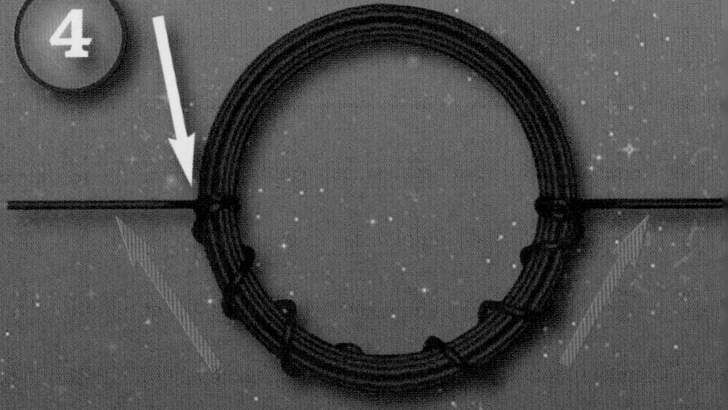

Be sure the two leads are well centered and straight.

5 Sand the top half of each lead.

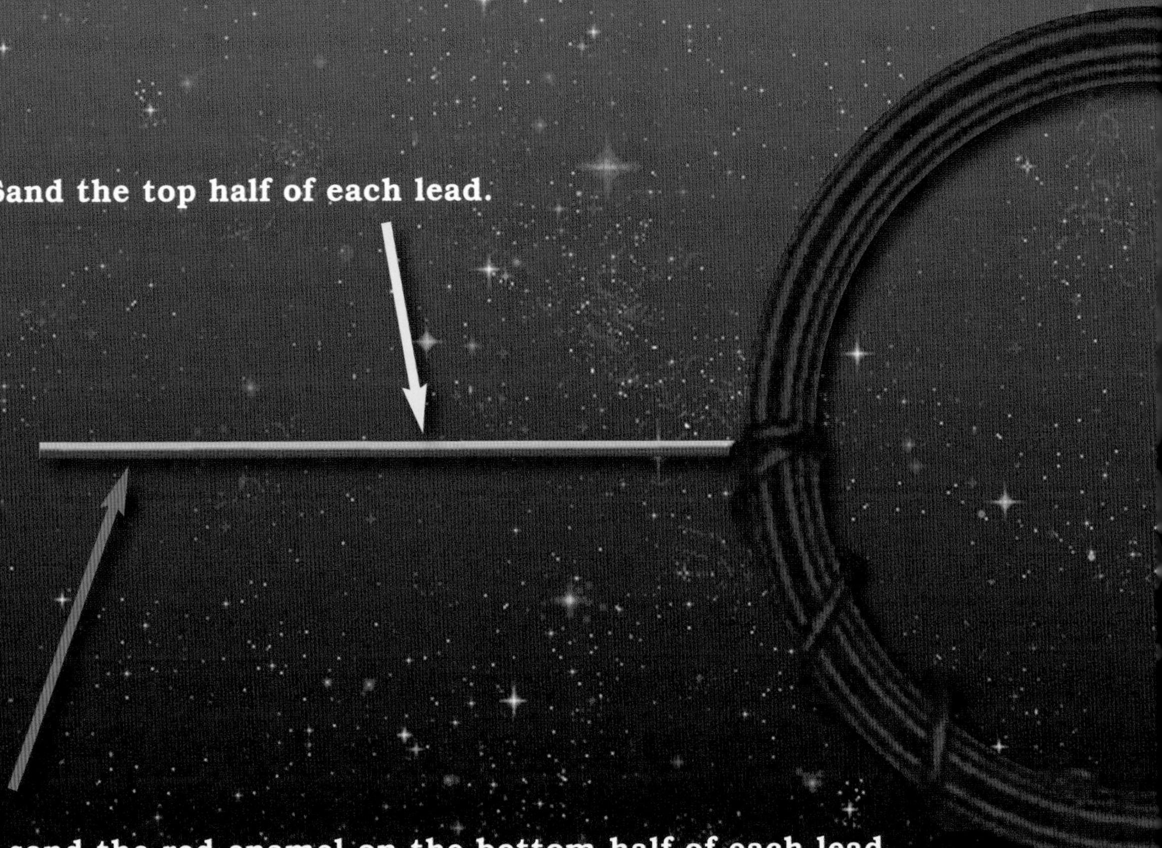

DO NOT sand the red enamel on the bottom half of each lead.

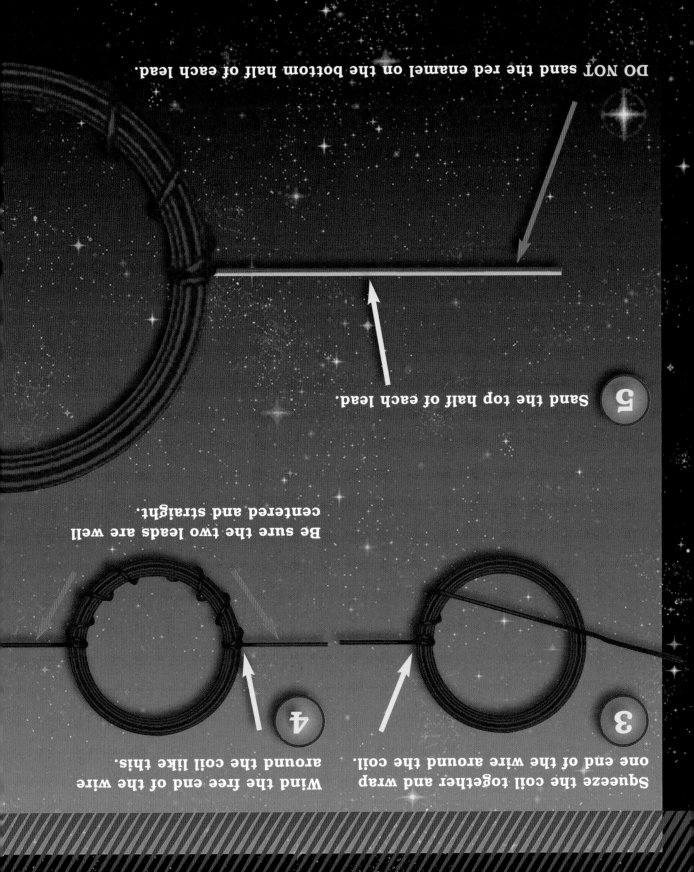

SANDING IS IMPORTANT: Be sure the paint is completely removed from the top half of each lead. The sanded top half should shine.

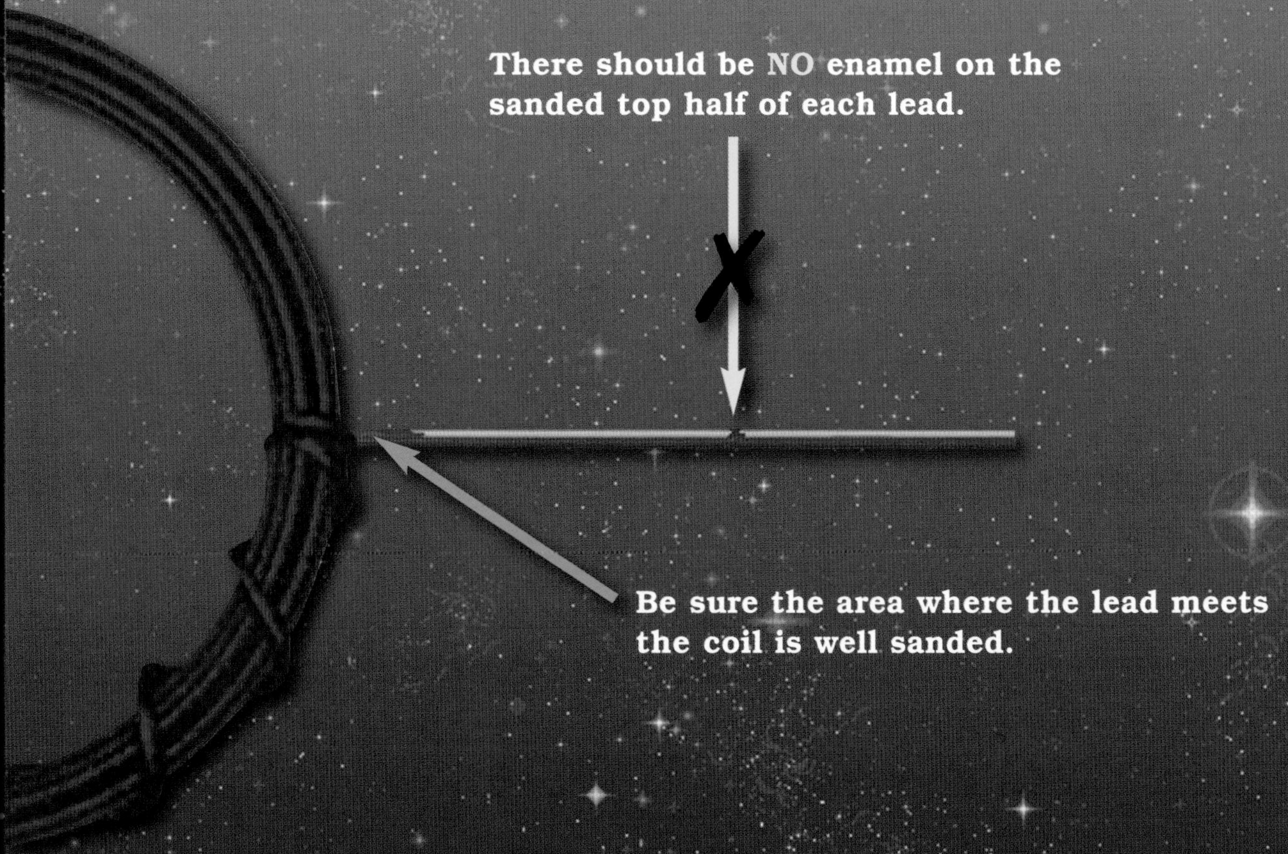
There should be NO enamel on the sanded top half of each lead.
Be sure the area where the lead meets the coil is well sanded.

6 Now that the coil is finished, find the 2 jumbo sized paperclips and unfold each one.

7 Connect the motor together like this.

Loop the battery band around a D-Cell battery.

You may need to give the coil a gentle tap or twist to get it started.

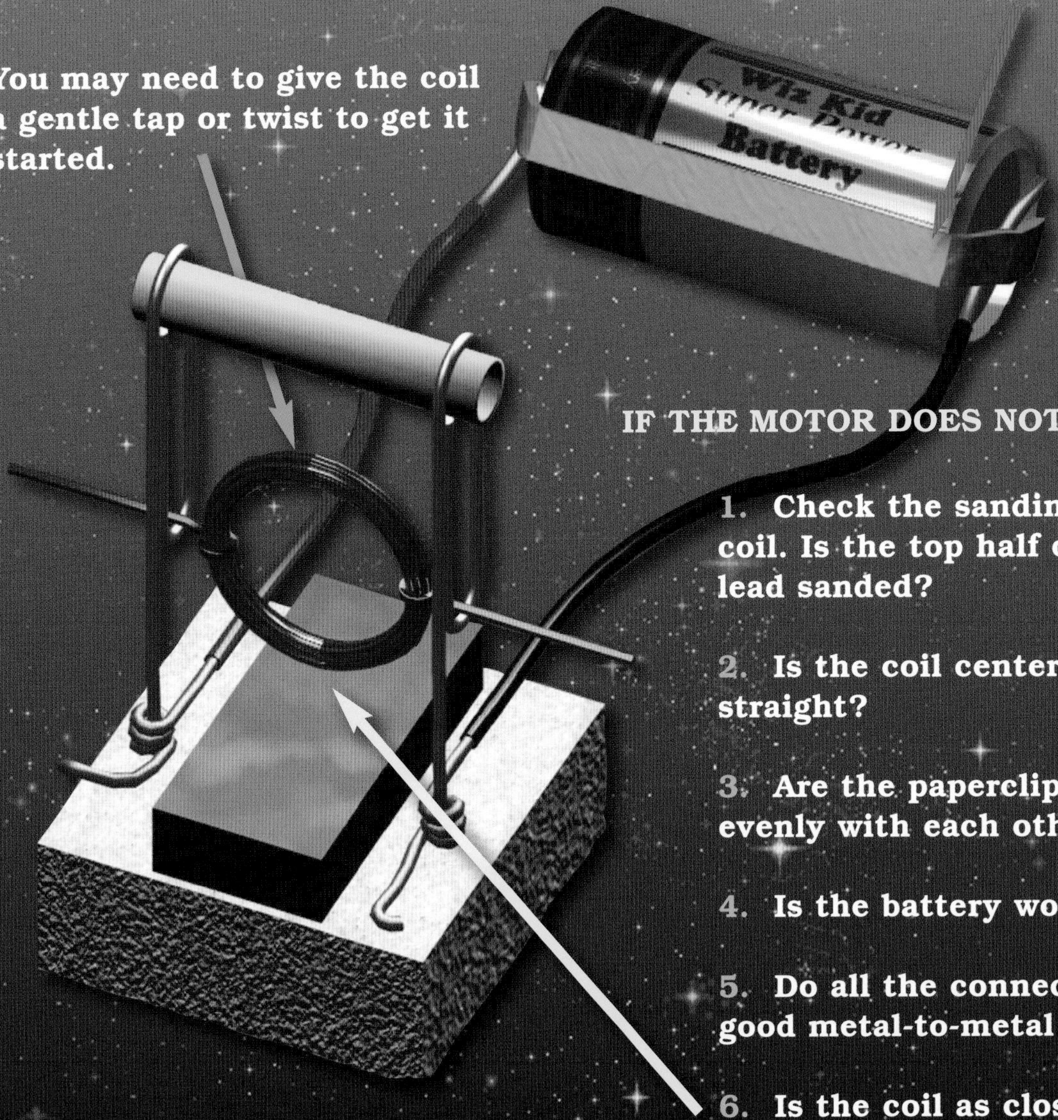
Wiz Kid
Battery

IF THE MOTOR DOES NOT WORK:

1. Check the sanding on the coil. Is the top half of each lead sanded?

2. Is the coil centered and straight?

3. Are the paperclips lined up evenly with each other?

4. Is the battery working?

5. Do all the connections have good metal-to-metal contact?

6. Is the coil as close to the magnet as possible without touching it?

7. If the bar magnet hits the paperclips, put a rubber band around the magnet and foam.

HOW THE MOTOR WORKS

This motor is actually a half-motor.

1

Half of the time the COIL is an electromagnet. This happens when the SANDED side of each lead is in contact with the paperclips.

To show this, put a compass near the coil. The compass needle should move and point at the coil. This shows that electricity is flowing through the COIL and that the COIL is an electromagnet.

2

When the enameled side of each lead is pressed against the paperclips, the coil is NOT an electromagnet.

You can see this for yourself by holding the enameled side of the coil's leads against the paperclips. No electricity will flow through the coil. The coil will NOT move the compass needle. The compass will point north.

This is how the ON again, OFF again of the coil creates motion. When the COIL is an electromagnet, it turns to line up with the bar magnet and the motion begins. Just as this happens, the enameled halves of the leads touch the paperclips. Electricity stops flowing through the coil. The electromagnet is OFF, but the coil is moving and it keeps on moving long enough for the cycle to repeat.

So half the time the coil is an electromagnet that moves to line up with the permanent magnet below and half the time the coil is just coasting. Hence you have made a HALF* motor.

* In a full motor the coil is never OFF.

The marvel of the motor is that a coil and a magnet turn electricity into

MOTION!

MORE THINGS TO TRY:

1. Connect a light bulb in a loop (in series) with the motor. What happens? Why?*
2. Does adding a second battery or using other magnets affect the motor?
3. How does changing the number of turns in the coil affect the motor?
4. You can turn the motor into a galvanometer with just a toothpick and sand paper. Go to www.ScienceWiz.com to learn how.

*The bulb flickers. This demonstrates the on-off pattern of the coil as it rotates.

Half the time, when the sanded sides of the coil are against the paperclips, electricity is flowing through the coil, as well as the bulb. The bulb is on. As electricity flows through the coil, it becomes an electromagnet. This electromagnet lines up with the bar magnet below. Remember, unlike poles attract and like poles repel. So the magnetic poles going through the coil's centre will turn the coil's face toward the permanent magnet.

Just as this is happening, the electricity turns off because the enameled half of the coil's leads do not conduct electricity. No electricity can flow through the coil or the bulb. The coil acts as a switch. When no electricity flows through the coil, the bulb is off. The coil stops being an electromagnet, but the coil just keeps moving. Why? Just like a child on roller skates, the coil keeps on rolling after an initial push. Once in motion the coil stays in motion.

A MOTOR IN REVERSE - A GENERATOR

With a motor you can generate electricity. Here's how.

1. Find the motor, the bulb holder with the bulb, and the string.

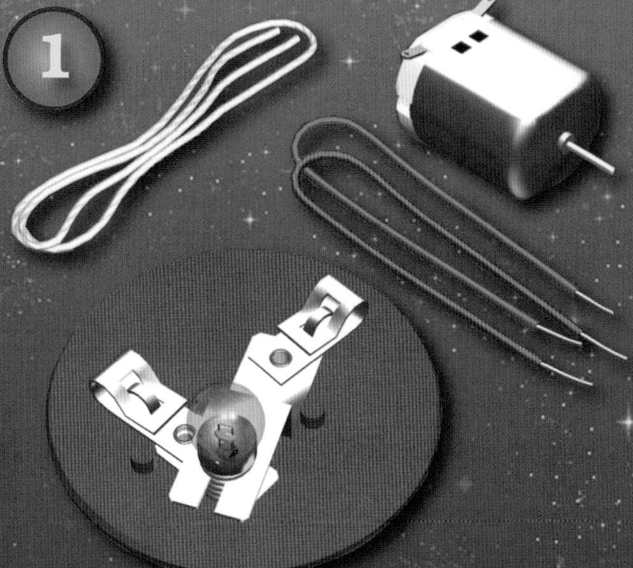

2. Connect the motor to the light bulb holder.

3. There are at least two ways to do the next step:

Method 1: The Coordinated Fingers Method:

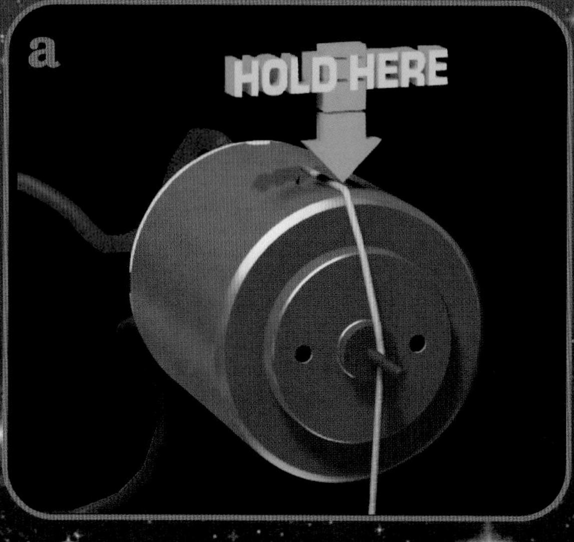
a
HOLD HERE

Hold the string in place or tie the string around the end of the shaft of the motor.

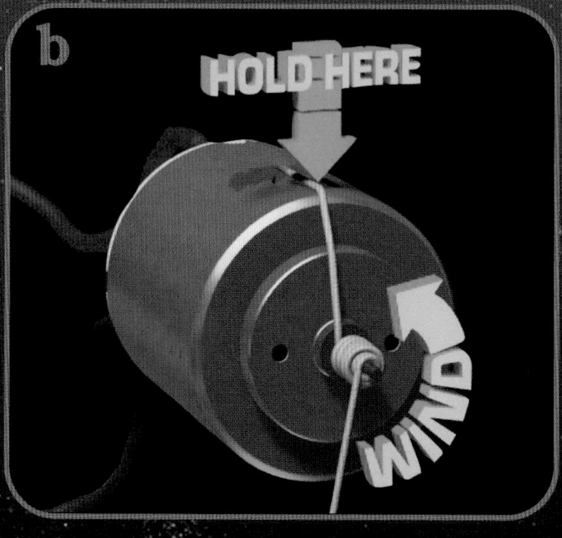
b
HOLD HERE
WIND

Wind the string around the shaft of the motor at least 20 times.

If your string is now on the shaft of the motor, go to Step 4.

If Method 1 was too difficult, this is a quicker way to get a result, but you may want to protect your tension finger with a bandage.

Method 2: The String Burn Method:

a
HOLD HERE
SHORT END

Wind the string once or twice around the shaft of the motor.

b
HOLD
TENSION
WIND
PULL
SHORT END

Use your bandage protected finger to hold tension on the long end of the string. You will pull the short end of the string.

No matter which method you used, you are ready for the next step.

4 While holding the motor in one hand, yank or pull quickly on the string as you would to start a lawn mower or to spin a top.

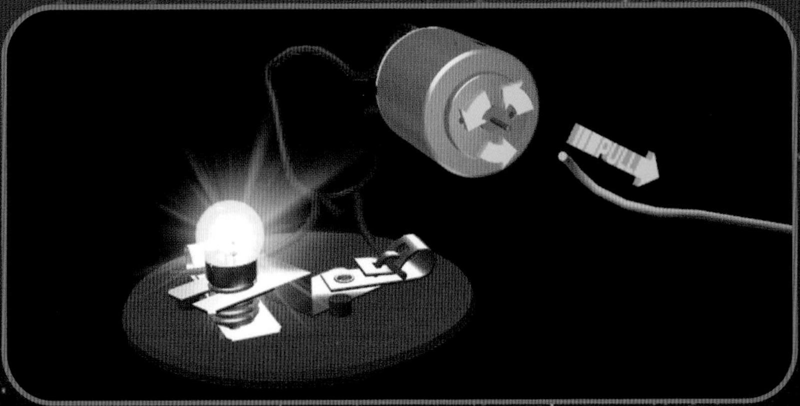
PULL

Watch the light. If it does not shine, try it until one of the methods works.

Go to www.ScienceWiz.com to build a bicycle-based generator.

HOW A GENERATOR WORKS

You have made ***ELECTRICITY*** by spinning the shaft of a motor.

You have turned an electric motor into a generator! Generators make electricity.

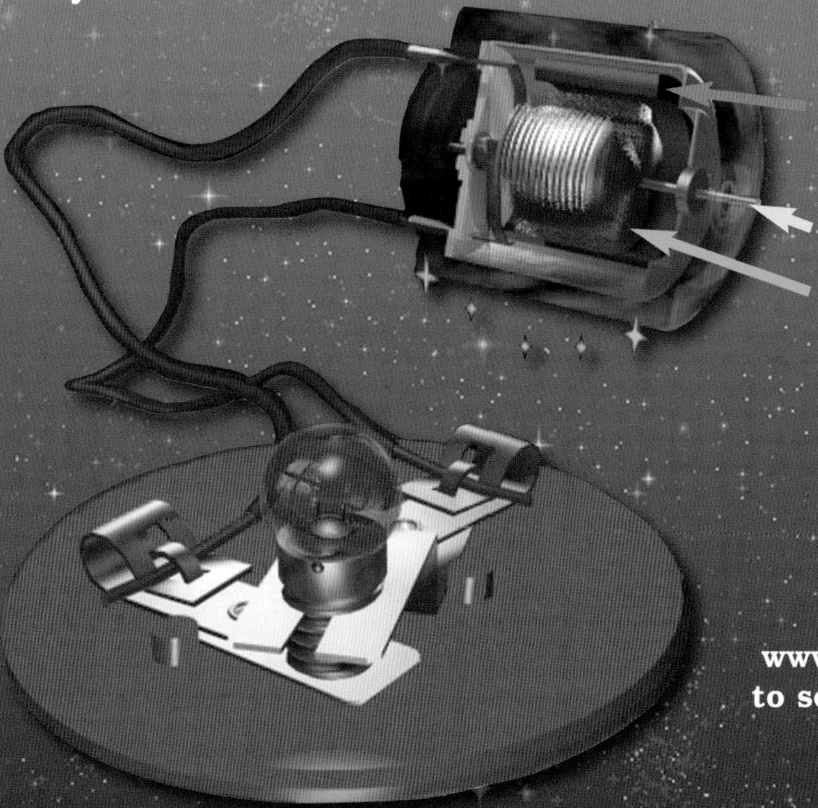

Inside the Motor:

permanent magnet (the second magnet is cut away)

shaft

coils

Go to www.ScienceWiz.com to see how this works.

HOW DOES THIS WORK?

When you spin the shaft of the motor, you are also spinning coils. Look at the picture to see how the shaft is attached to the coils inside the metal casing. If you took apart the motor, you would see three coils and two permanent magnets surrounding the coils.

Here is the WONDER: By moving a coil near a magnet, electrical current will flow in the coil!* By spinning the coil in the motor, you generated enough electricity to light a bulb!

* Electrons in the wire move in response to the wire's motion through the magnets' magnetic field. Moving any loop of wire near a magnetic field generates electrical current in the wire.

Most electrical power plants generate electricity in much the same way. Power plants generate electricity by moving large coils of wire near a magnet or by moving a magnet near a coil. The movement of coils of wire through a magnetic field generates *electricity* in the wire!

People have invented many ways to spin MAGNETS or COILS to make electricity.

They use falling water to turn magnets! This is called hydroelectric power. Hoover Dam on the Colorado River in the United States, the power station at Niagara Falls on Canada's border, Australia's Snowy Mountains Scheme, and China's Three Gorges Dam all make use of hydroelectric power. These power plants harness the powerful force of falling water to spin magnets which generate enough electricity for whole regions.

Waterfalls are used to spin huge magnets. The spinning magnets generate *electricity* in surrounding wires.

Another way to spin magnets or coils is with steam from boiling water. The force of steam shooting out from a pipe at high speed is so powerful that it can turn the blades on a wheel connected to magnets. The spinning of these huge magnets near coiled wire generates electrical current in the wire. Can you think of other ways to generate steam to turn a magnet?

BUILDING A TELEGRAPH

You can use a COIL to make a clicking telegraph. You will need:

Connecting Wire
Battery Band
Magnet Wire
(70" or 180 cm)
1 Piece of Cardboard
SWITCH
16d Nail
D-Cell Battery
2 Paper Fasteners
Electro Wiz
Super Power
Battery
Packing Foam
(2 X 2" or 5 X 5 cm square
and 3/4" or 2 cm deep)
Sandpaper
(Medium Grit)
2 Jumbo Sized
Paperclips

1. Begin the telegraph by winding the magnet wire around the nail 90 to 100 times.

2. Stick the nail into the foam base to keep the wire from sliding off the nail.

3. Be sure to leave about 8-10" or 20-25 cm of wire free at each end of the coil.

4. Use sandpaper to completely remove 1" or 2.5 cm. of enamel from both ends of the wire.

By connecting this coil to a battery you can make an

ELECTROMAGNET.

To test this, try picking up some paperclips.

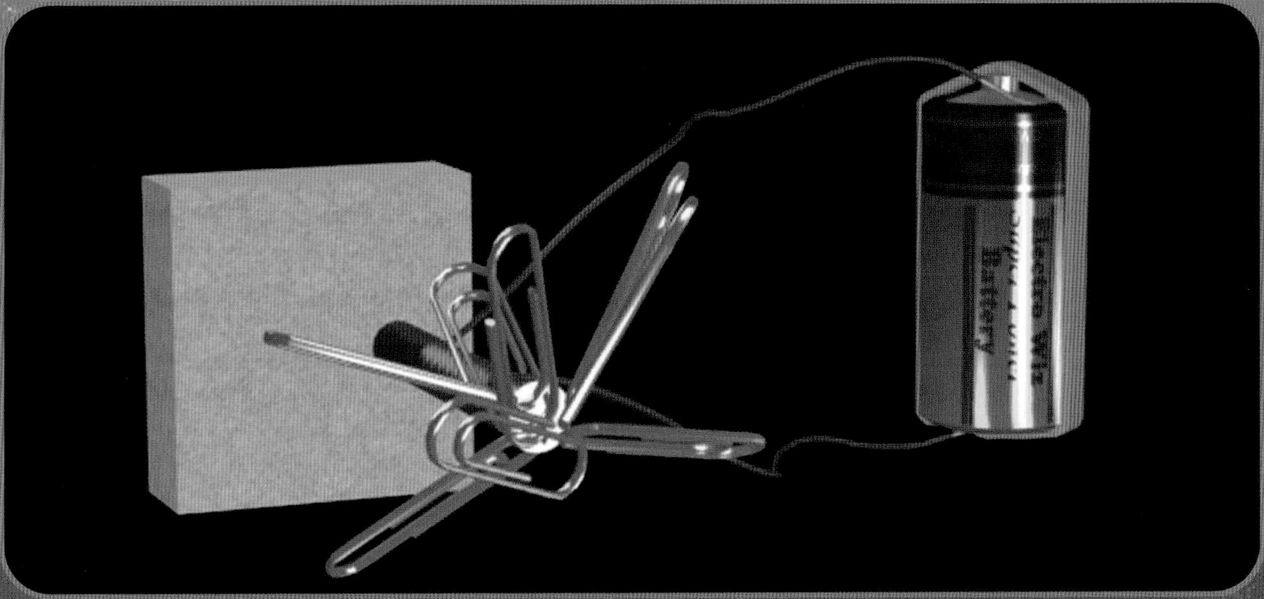
Electro Wiz
Battery

When you finish using the electromagnet, be sure to disconnect the battery.

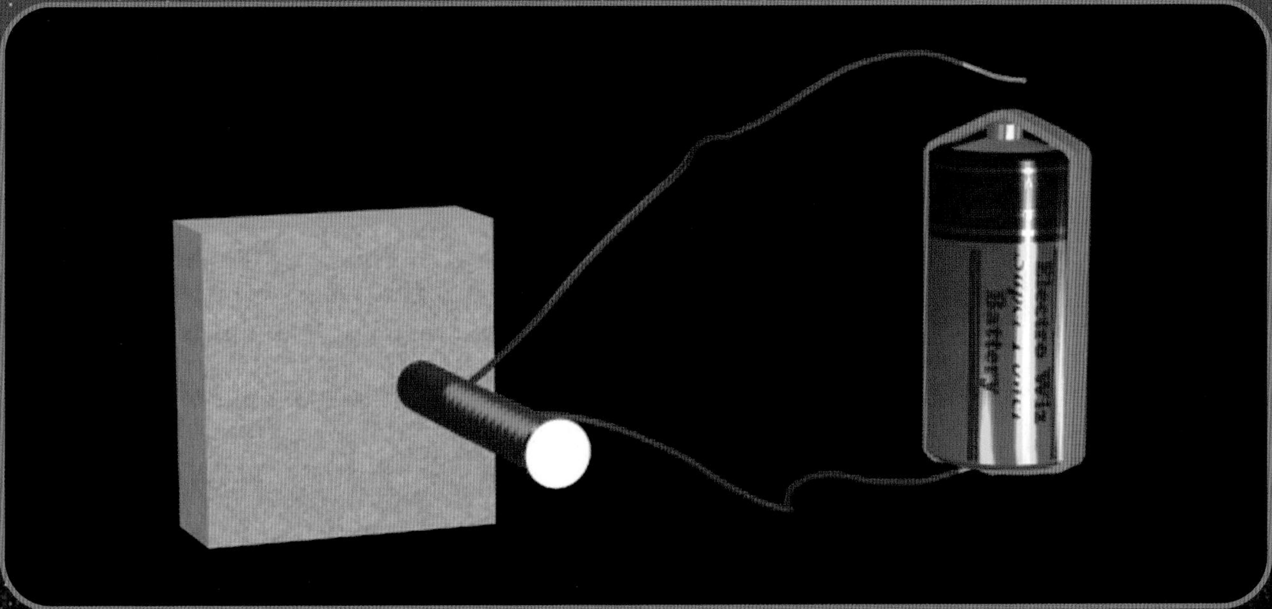
Electro Wiz
Battery

To turn the electromagnet into a telegraph, you will need to make a telegraph clicker and a telegraph key.

Here's how to make a telegraph clicker with a paperclip.

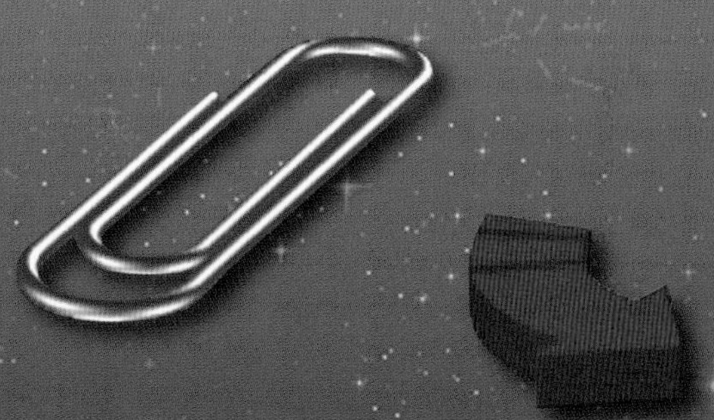

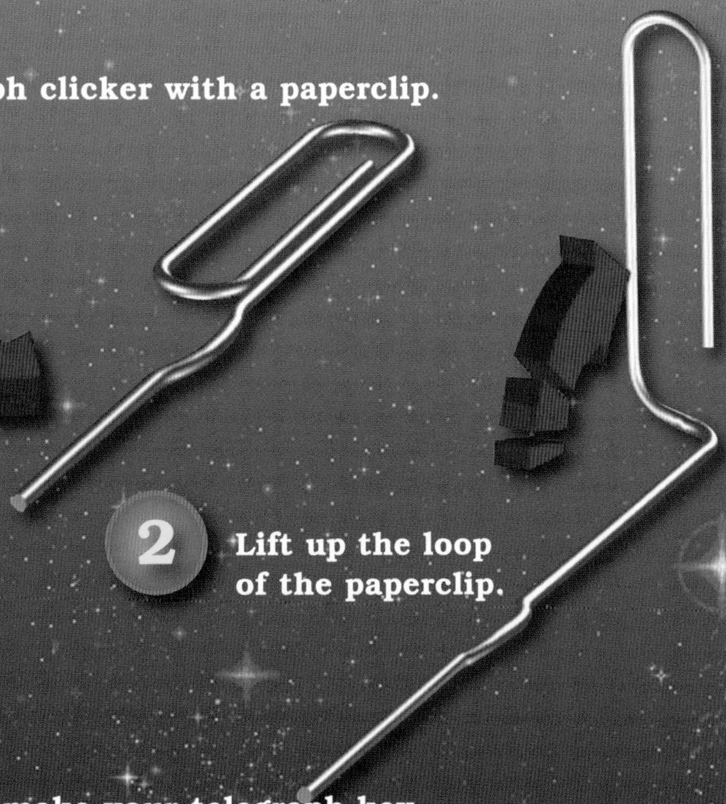

1 Unfold the paperclip.

2 Lift up the loop of the paperclip.

Now set the clicker aside and make your telegraph key.

To make a telegraph key to send messages, you will need a second paperclip, two paper fasteners and a piece of cardboard.

1 Pull up the outer loop of the paperclip.

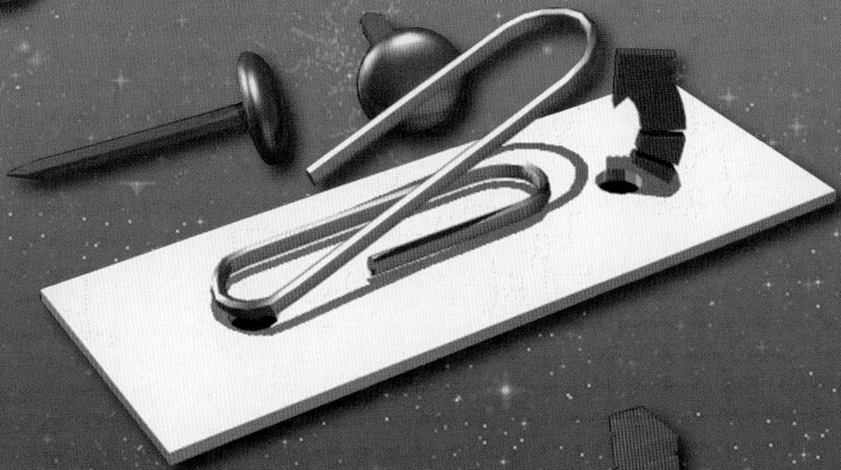

2 Push the two paper fasteners through the holes of the cardboard, with the paperclip positioned like this.

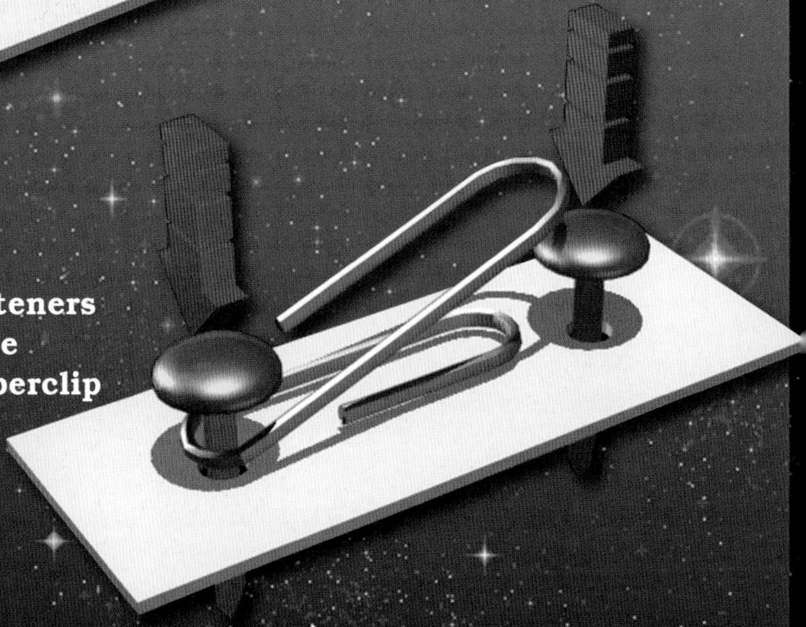

3 Pull open the two paper fasteners.

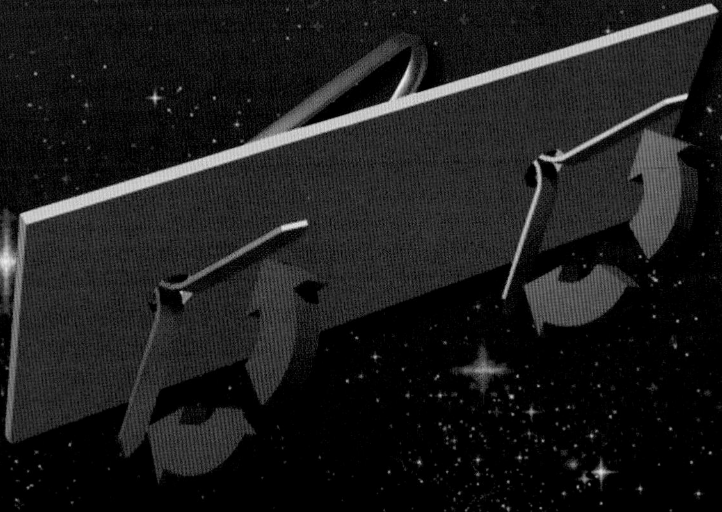

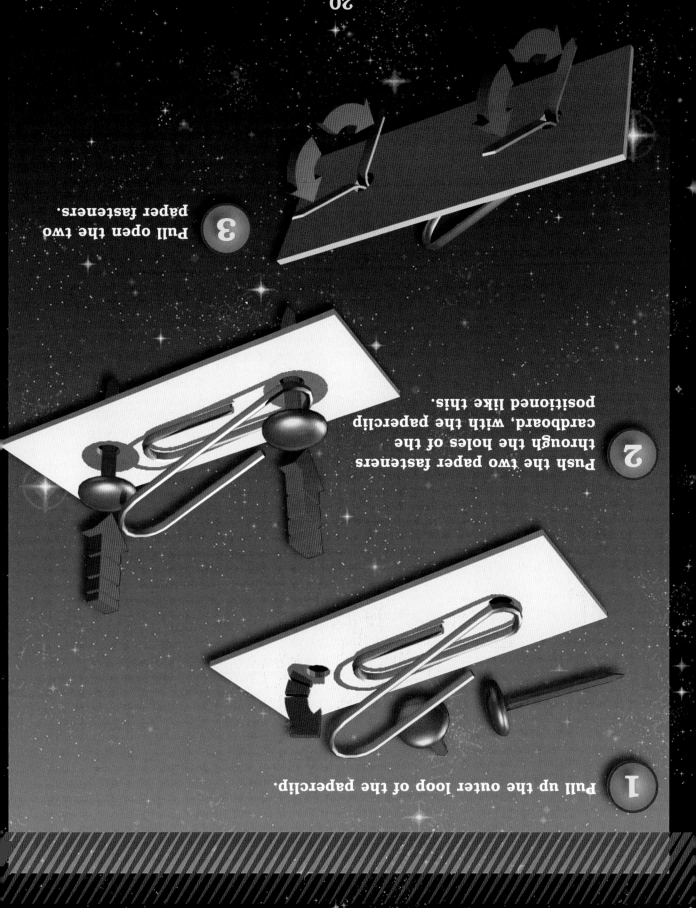

MAKING THE TELEGRAPH WORK

Put the telegraph together like this.

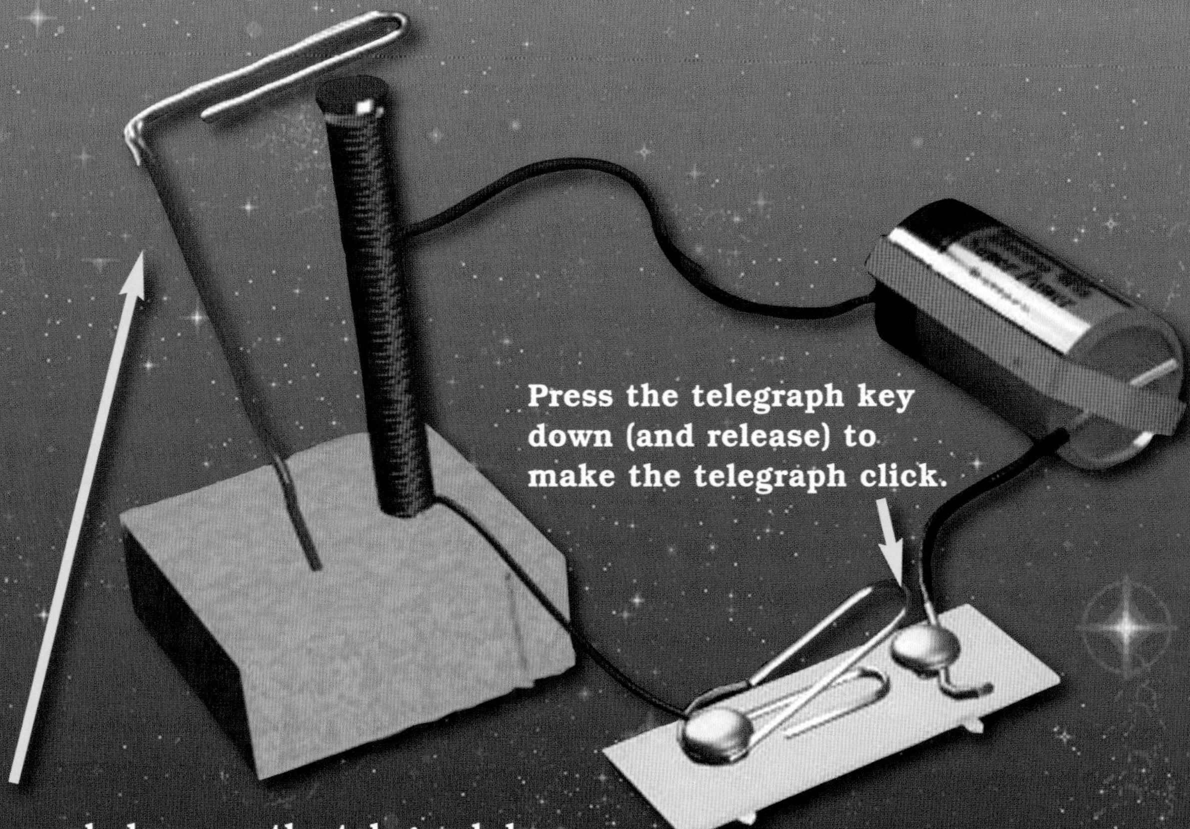
Press the telegraph key
down (and release) to
make the telegraph click.

As you push down on the telegraph key, the telegraph clicker should make a clicking sound as it hits the nail.

If the clicker is too far from the nail, it will wiggle but it will not click.

If the clicker is too close, it will just stick.

If the clicker does not move at all, check your battery and connections.

You can use your telegraph to send coded messages to a friend.

Do you know what metal is in the telegraph clicker?
Hint: What metal is attracted to a magnet? IRON

HOW THE TELEGRAPH WORKS

You may want to attach your telegraph to cardboard.

1. Cut two slits about 2" or 5 cm apart in a piece of cardboard.

2. Stretch the rubber band around the slits and the telegraph base, like this.

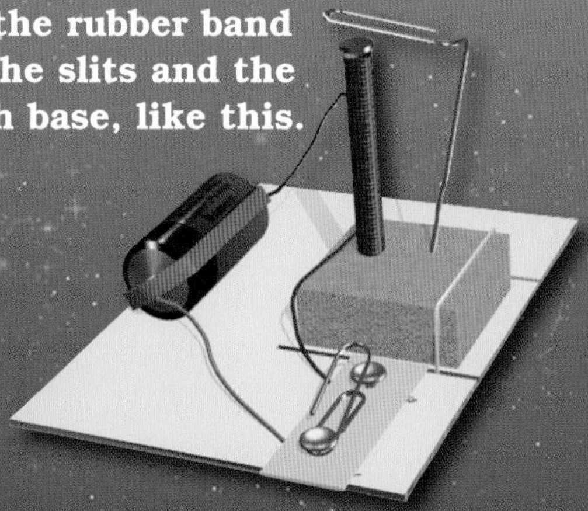

In the trick of the telegraph, the coil and the paperclip turn electricity into SOUND.

How does the telegraph work? When you push down on the telegraph key, you are closing the loop, allowing electrical current to flow through the coil. The COIL becomes an ELECTROMAGNET. The nail strengthens the magnetic field. The iron in the paperclip clicker above the nail is attracted to the electromagnet. The clicker makes a clicking sound as it strikes the nail.

Sound!!!

With a telegraph you can send coded messages to a friend.

SENDING CODED MESSAGES

Samuel Morse made a simple code for the alphabet. This code was used to send messages over long distances with the telegraph.

There were two types of signals in Morse Code.

DI - a short key stroke, written as a short dot (.)
DA - a long key stroke, written as a long line (_)

Close and open the switch on the telegraph to make a quick short DI and the longer DA sound. Can you make your telegraph produce two different sounds?

One way to call for help is SOS. Ships at sea send this message to other ships when they need HELP. SOS has come to stand for Save Our Ship.

You can send your own SOS message in Morse Code. Look in the chart below for the two letters S and O. (S is ... and O is _ _ _). Send the SOS message with your telegraph.

DI DI DI DA DA DA DI DI DI

THE INTERNATIONAL MORSE CODE

A . _	I . .	Q _ _ . _	Y _ . _ _	6 _ . . . .
B _ . . .	J . _ _ _	R . _ .	Z _ _ . .	7 _ _ . . .
C _ . _ .	K _ . _	S . . .	0 _ _ _ _ _	8 _ _ _ . .
D _ . .	L . _ . .	T _	1 . _ _ _ _	9 _ _ _ _ .
E .	M _ _	U . . _	2 . . _ _ _	, _ _ . . _ _
F . . _ .	N _ .	V . . . _	3 . . . _ _	. . _ . _ . _
G _ _ .	O _ _ _	W . _ _	4 . . . . _	OVER _ . _
H	P . _ _ .	X _ . . _	5	OUT . _ . _ .

With a telegraph you can send coded messages almost as fast as the speed of light.

How far can you send these messages? Would you guess a few feet, meters, next door, across town or around the world?

MAKING A RELAY

How far can you send a telegraphic message? A single battery will send electrical current down a length of wire to a telegraphic clicker. Depending on the voltage of the battery, this distance can be as far as 50 to 100 miles or 80 to 160 km, but there is a limit. After that limit the telegraph will not work.

By adding relays you can increase the distance from across town to around the world.

Here's how to make a RELAY.

1. Connect the first telegraph together as you did before.
2. Build a second telegraph and connect it to the battery.

   Do not make a switch for the second telegraph.

3. Connect the battery here.

4. Connect the second telegraph to the first, as shown.

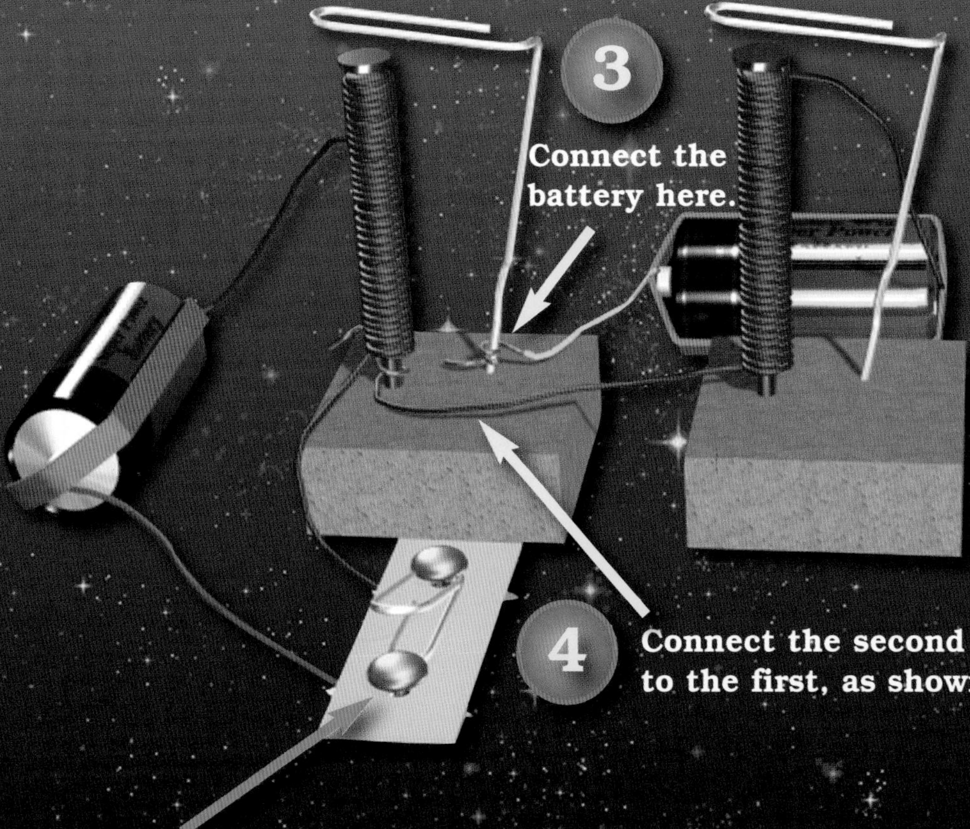

This telegraph key controls the clicking of both telegraphs.
You just made a RELAY. A relay is an electrically controlled switch.

BUILDING A RADIO

With a COIL you can build a working radio.

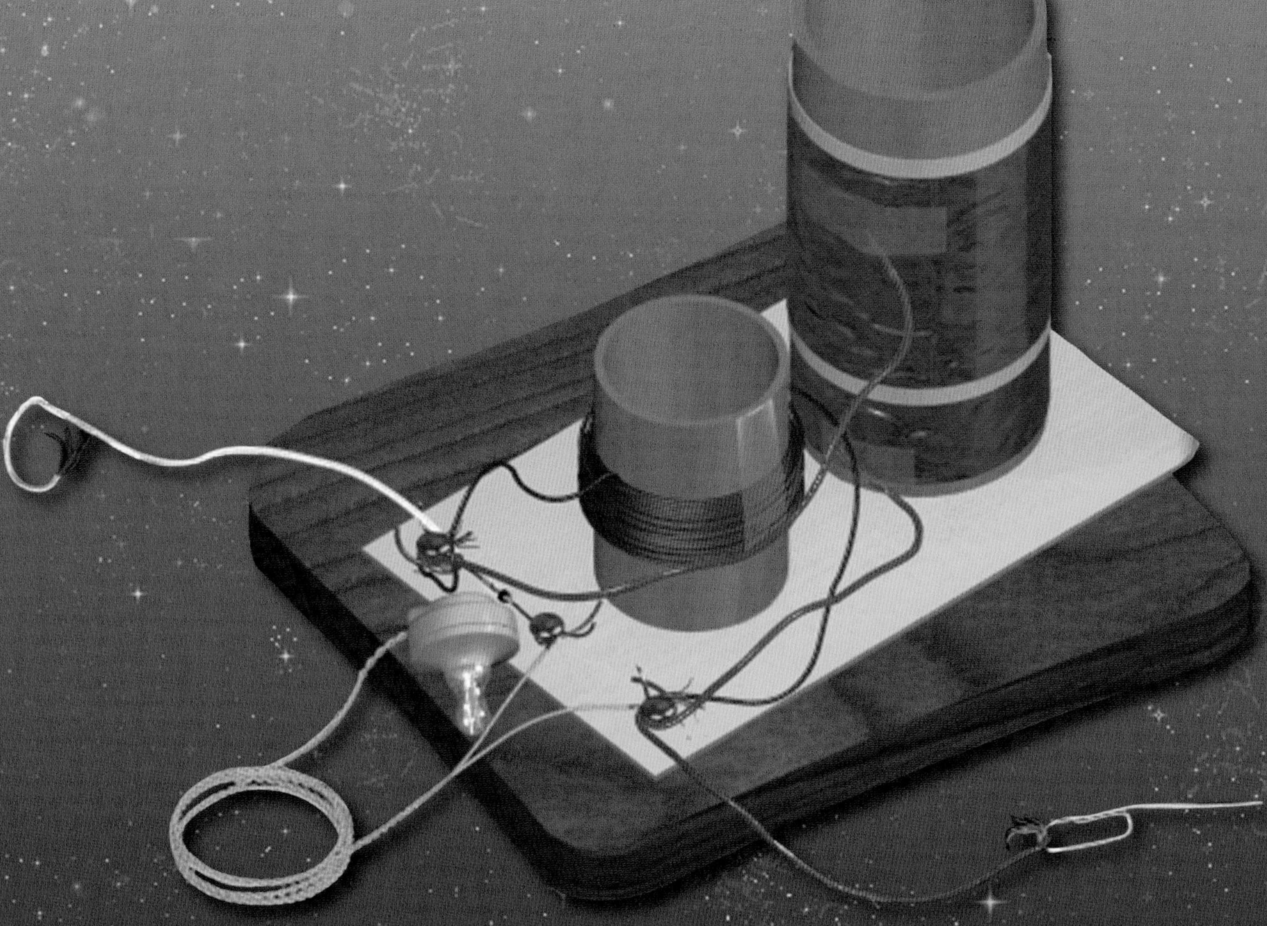

These are the tools you will need to build the radio.

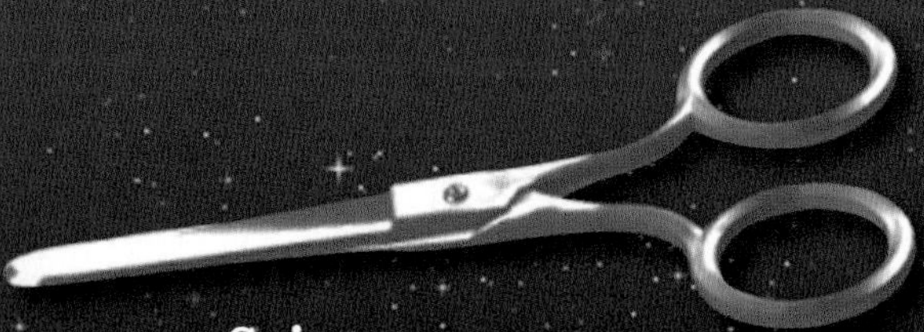

Scissors

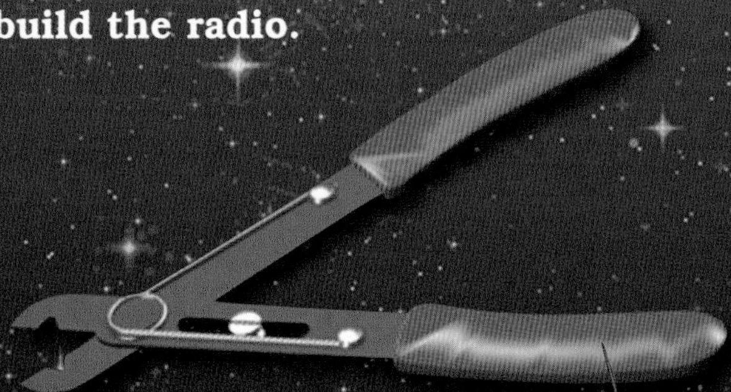

Wire Strippers
(the end of the wires in the kit are pre-stripped so you may not need these)

The wires, earphone and Schottky diode you will need are in the kit.

Schottky Diode
Connecting Wire
Magnet Wire on a spool
Earphone

These are common household items that you will need to build your radio. The sandpaper and paper fasteners are in the kit.

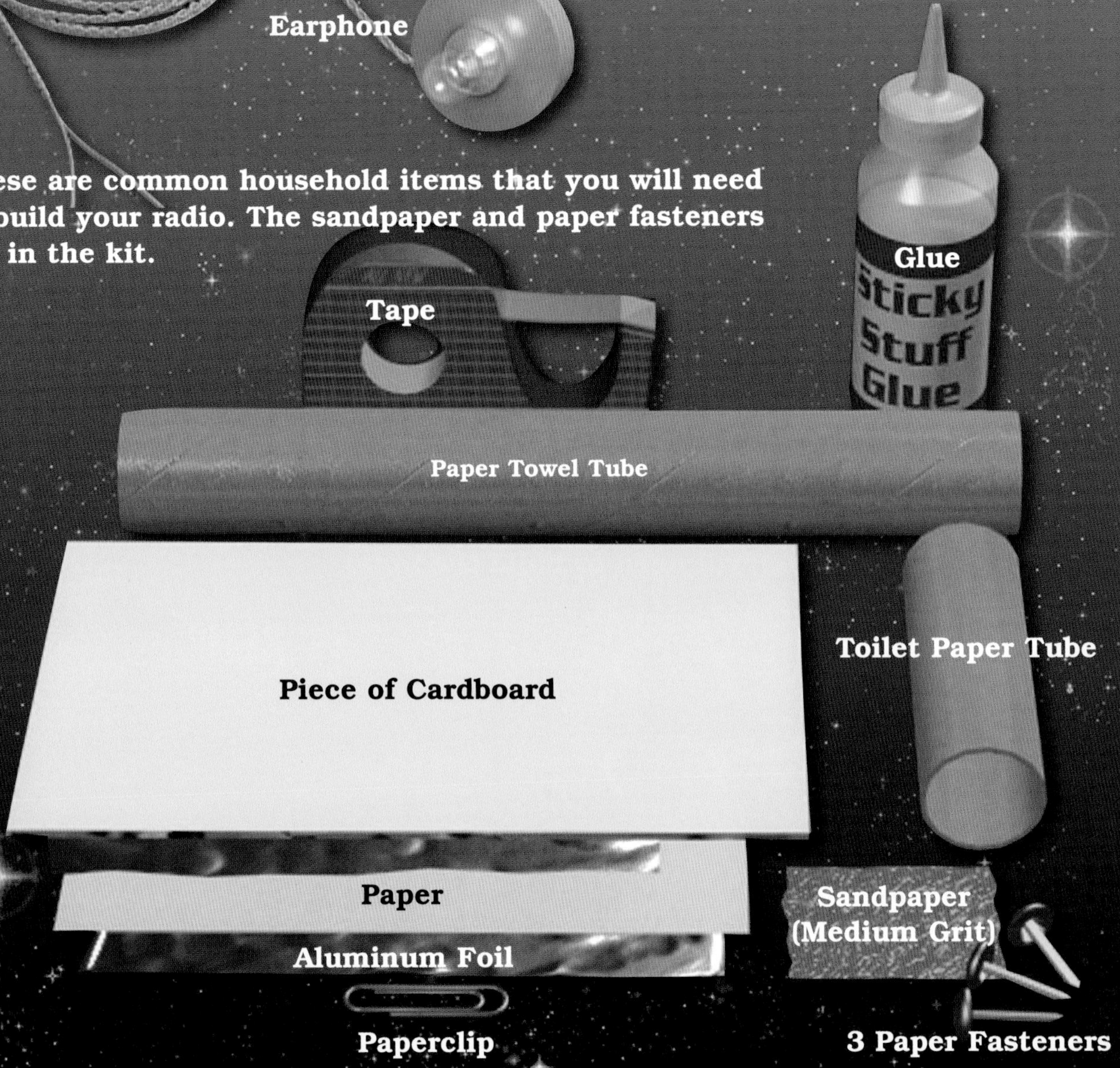
Glue
Sticky Stuff Glue
Tape
Paper Towel Tube
Piece of Cardboard
Toilet Paper Tube
Paper
Sandpaper (Medium Grit)
Aluminum Foil
Paperclip
3 Paper Fasteners

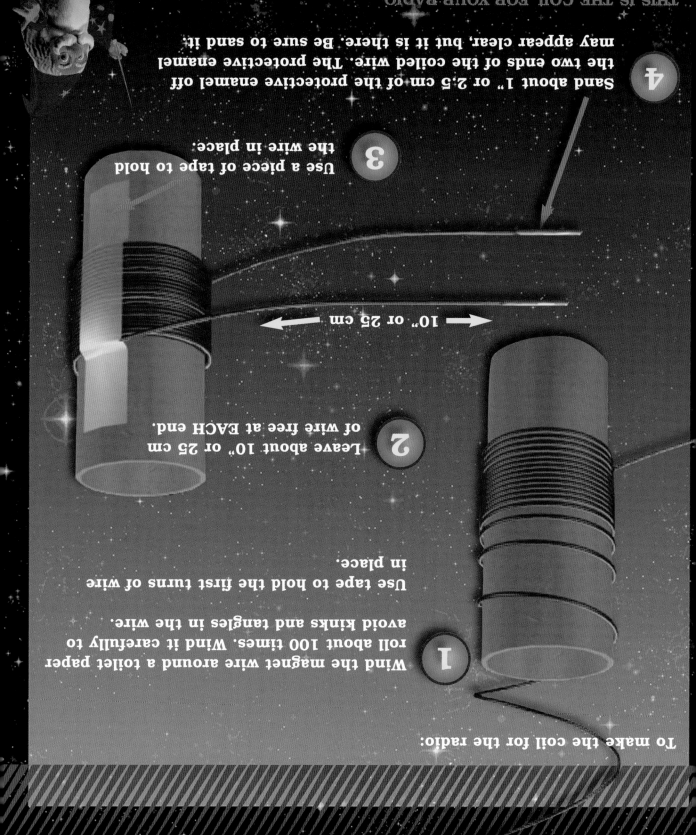

To make the coil for the radio:

1 Wind the magnet wire around a toilet paper roll about 100 times. Wind it carefully to avoid kinks and tangles in the wire.

Use tape to hold the first turns of wire in place.

2 Leave about 10” or 25 cm of wire free at EACH end.

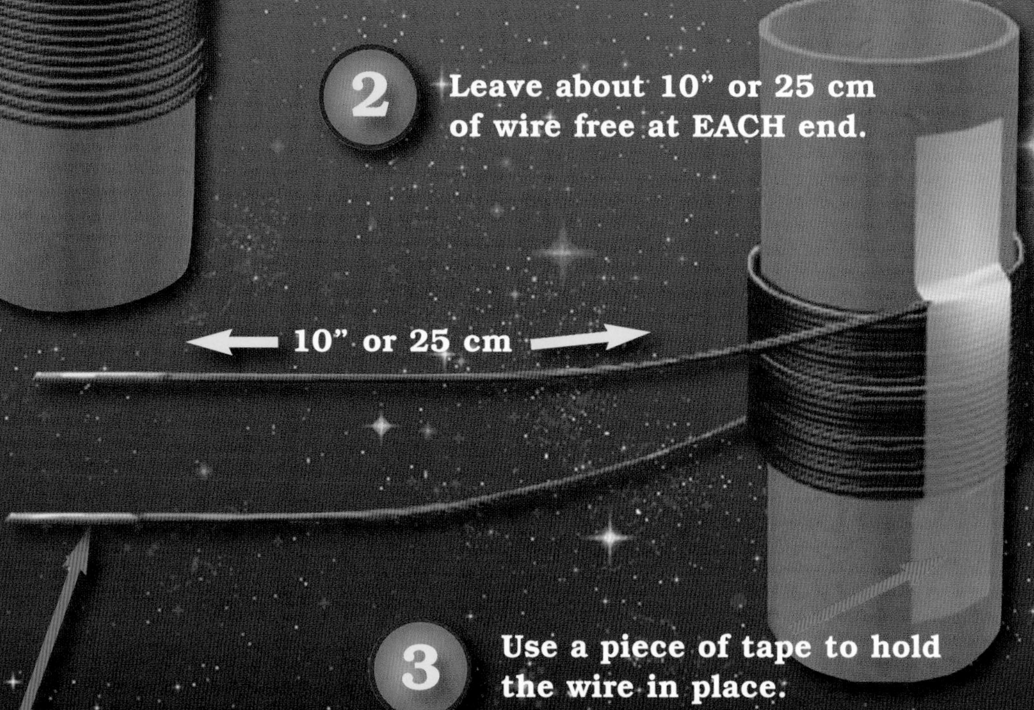
10” or 25 cm

3 Use a piece of tape to hold the wire in place.

4 Sand about 1” or 2.5 cm of the protective enamel off the two ends of the coiled wire. The protective enamel may appear clear, but it is there. Be sure to sand it.

THIS IS THE COIL FOR YOUR RADIO.

To make a variable capacitor:

1. Find an empty paper towel tube.

2. Cut a length of foil to fit around the tube. The foil should be half the height of the tube and be free of wrinkles. Use your fingernail to smooth the foil.

3. Tape the foil to the tube.

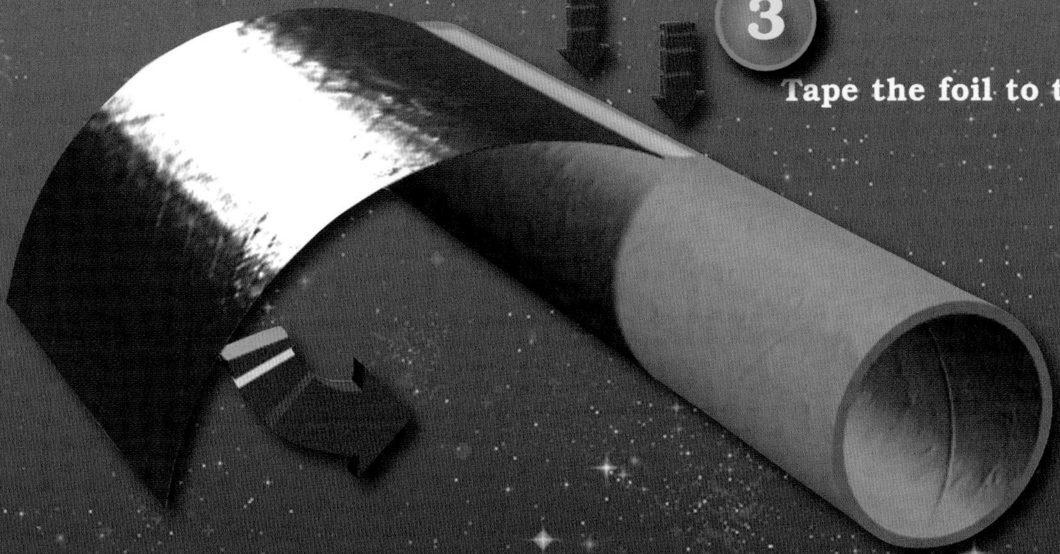

4. Wrap the foil around the tube.

5. Overlap the foil and tape it.

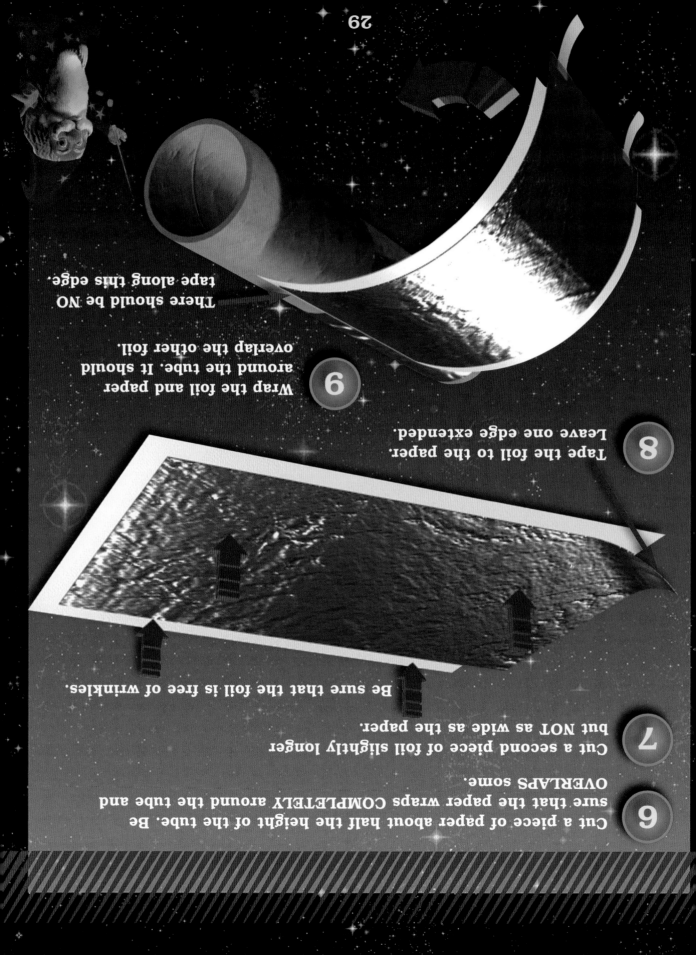

6. Cut a piece of paper about half the height of the tube. Be sure that the paper wraps COMPLETELY around the tube and OVERLAPS some.

7. Cut a second piece of foil slightly longer but NOT as wide as the paper.

Be sure that the foil is free of wrinkles.

8. Tape the foil to the paper. Leave one edge extended.

9. Wrap the foil and paper around the tube. It should overlap the other foil.

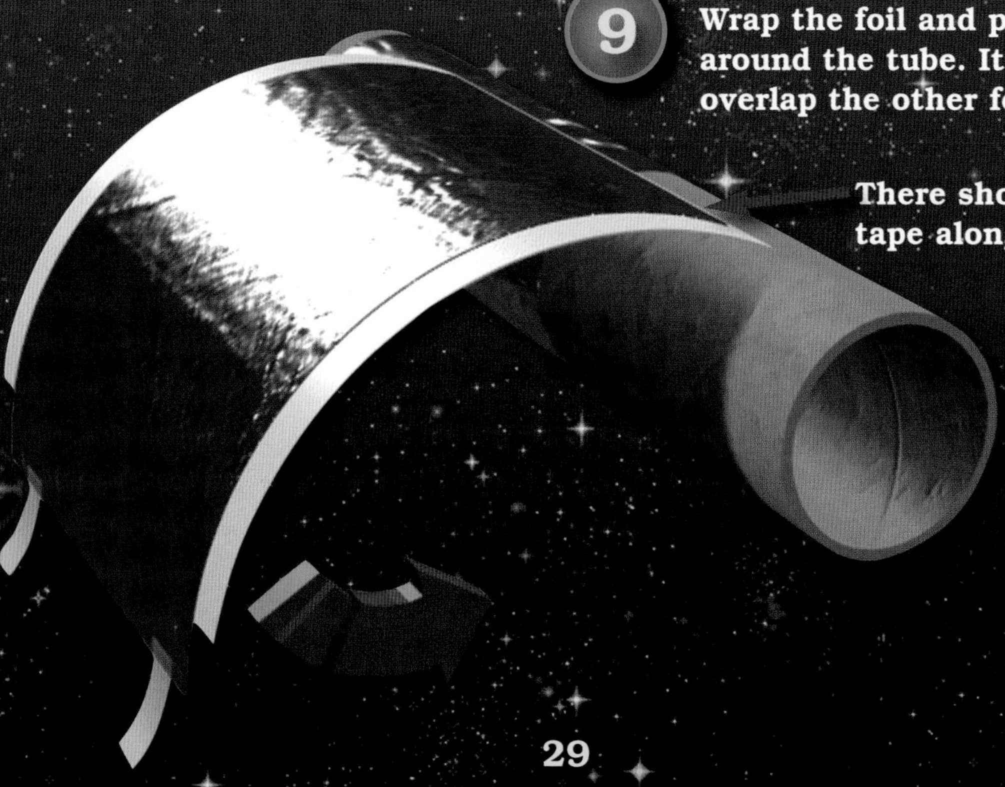
There should be NO tape along this edge.

10 Tape this paper/foil sheet so that it can slide up and down freely over the bottom foil.

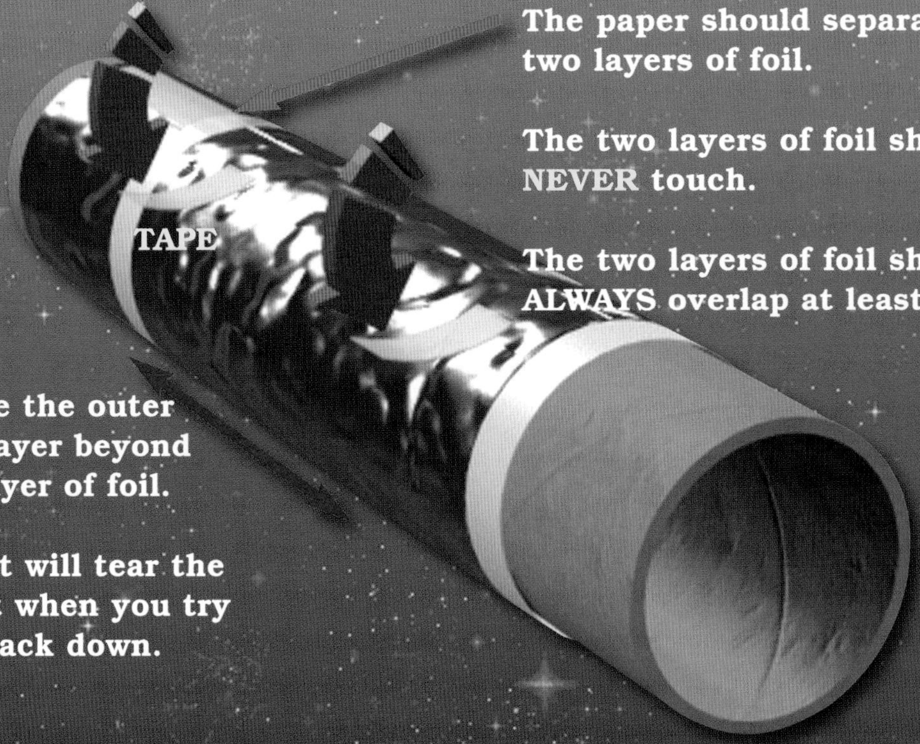
TAPE

The paper should separate the two layers of foil.

The two layers of foil should NEVER touch.

The two layers of foil should ALWAYS overlap at least SOME.

NEVER slide the outer paper/foil layer beyond the inner layer of foil.

If you DO, it will tear the foil below it when you try to slide it back down.

YOU HAVE JUST MADE A VARIABLE CAPACITOR.

Build a cardboard base to complete the radio. The base should be about 6" by 8" or 15 by 20 cm. or larger.

1 Make three holes about 1 inch or 2.5 cm apart along the shorter edge of the piece of cardboard.

2 Insert a paper fastener through each hole.

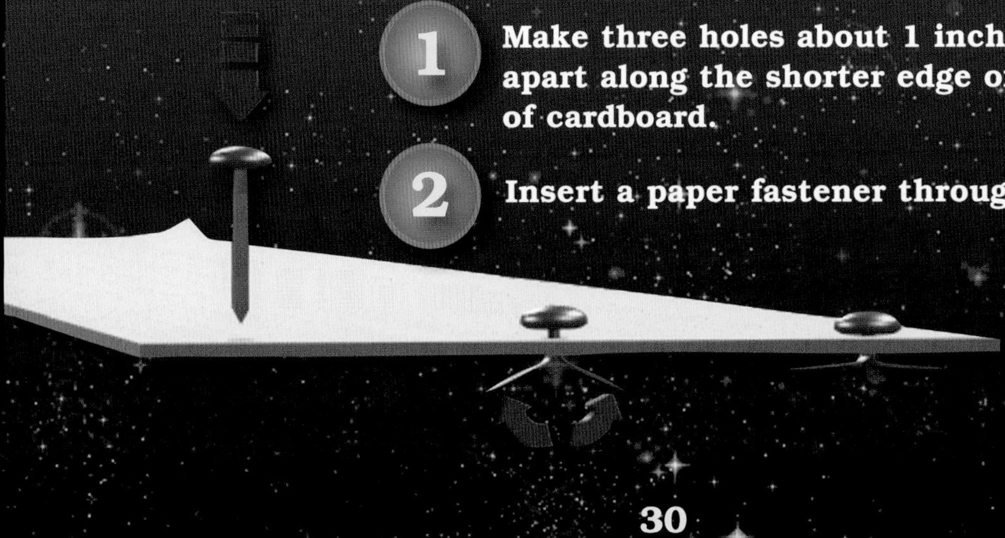

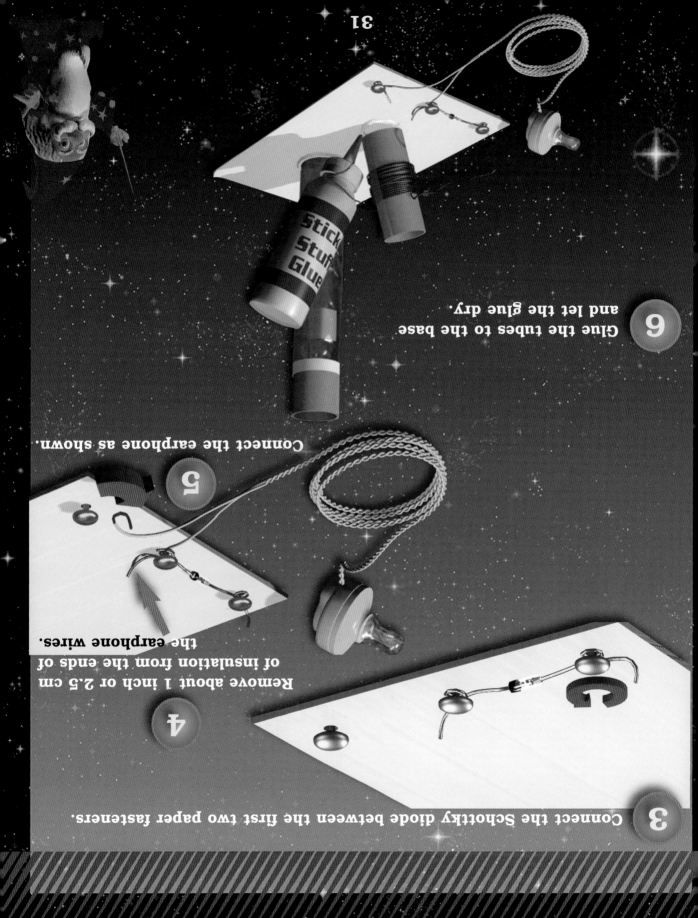

3 Connect the Schottky diode between the first two paper fasteners.

4 Remove about 1 inch or 2.5 cm of insulation from the ends of the earphone wires.

5 Connect the earphone as shown.

6 Glue the tubes to the base and let the glue dry.

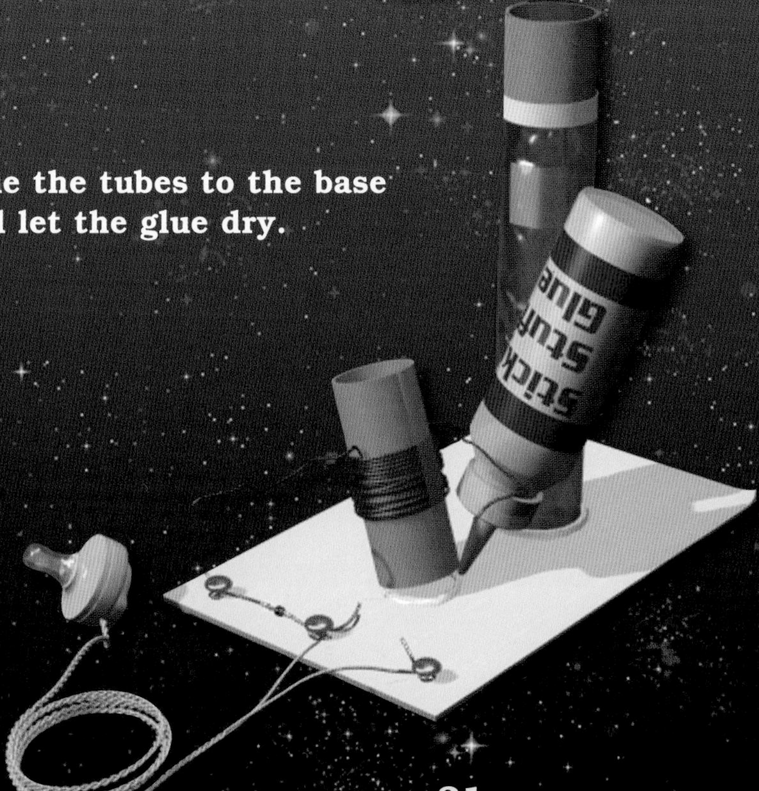
Stick
Stuff
Glue

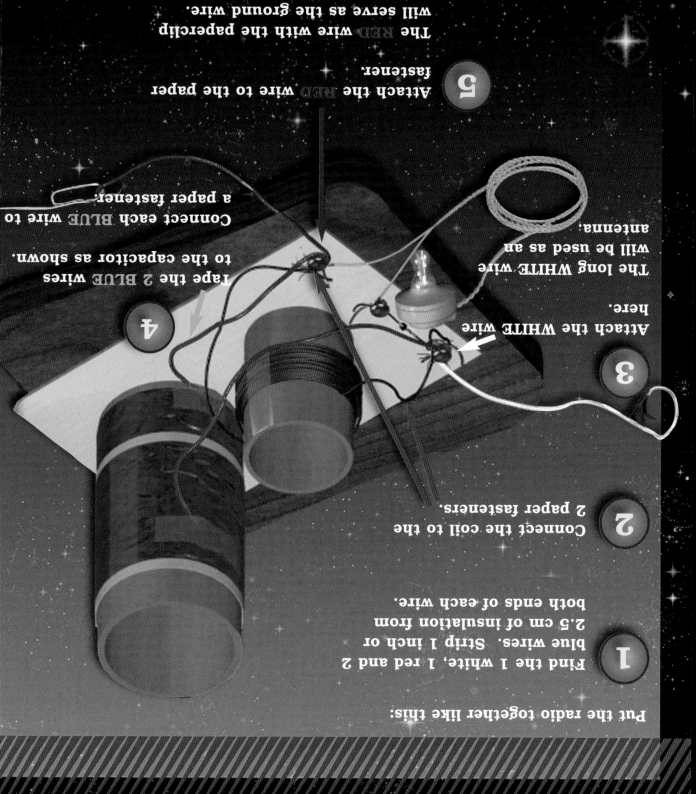

Put the radio together like this:

1 Find the 1 white, 1 red and 2 blue wires. Strip 1 inch or 2.5 cm of insulation from both ends of each wire.

2 Connect the coil to the 2 paper fasteners.

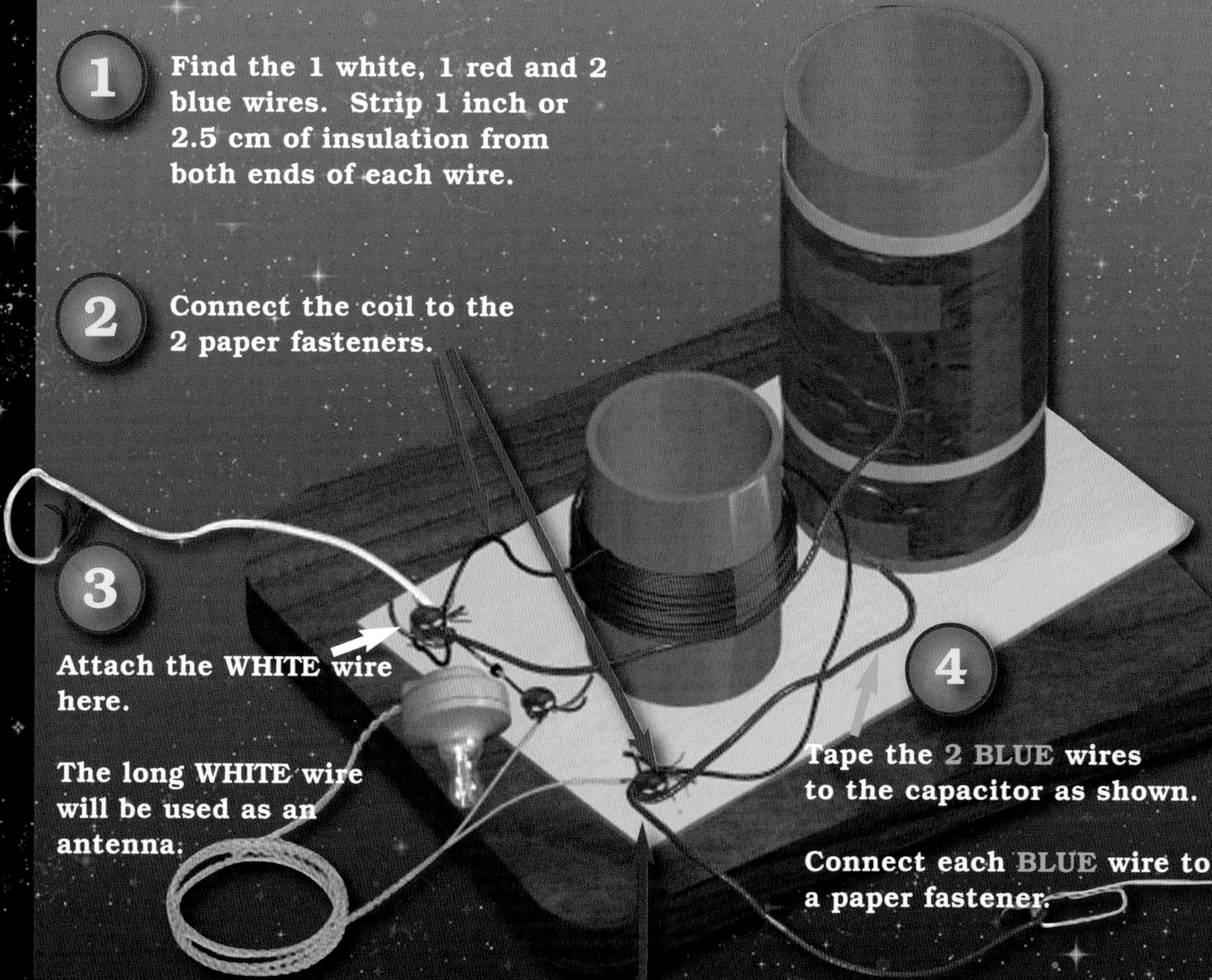

3 Attach the WHITE wire here.

The long WHITE wire will be used as an antenna.

4 Tape the 2 BLUE wires to the capacitor as shown.

Connect each BLUE wire to a paper fastener.

5 Attach the RED wire to the paper fastener.

The RED wire with the paperclip will serve as the ground wire.

First find a good antenna. If you live near a radio station, you can be your own antenna. Just hold the stripped end of the antenna wire with you fingers.

1. Otherwise, you can connect your white antenna wire to any large metal object. The antenna must not touch the ground directly or be connected by a conductor to the ground. Try a stretch of insulated wire, a television antenna or the spokes on the wheel of a bicycle. There must be good metal-to-metal contact between the antenna and the wire from your radio.

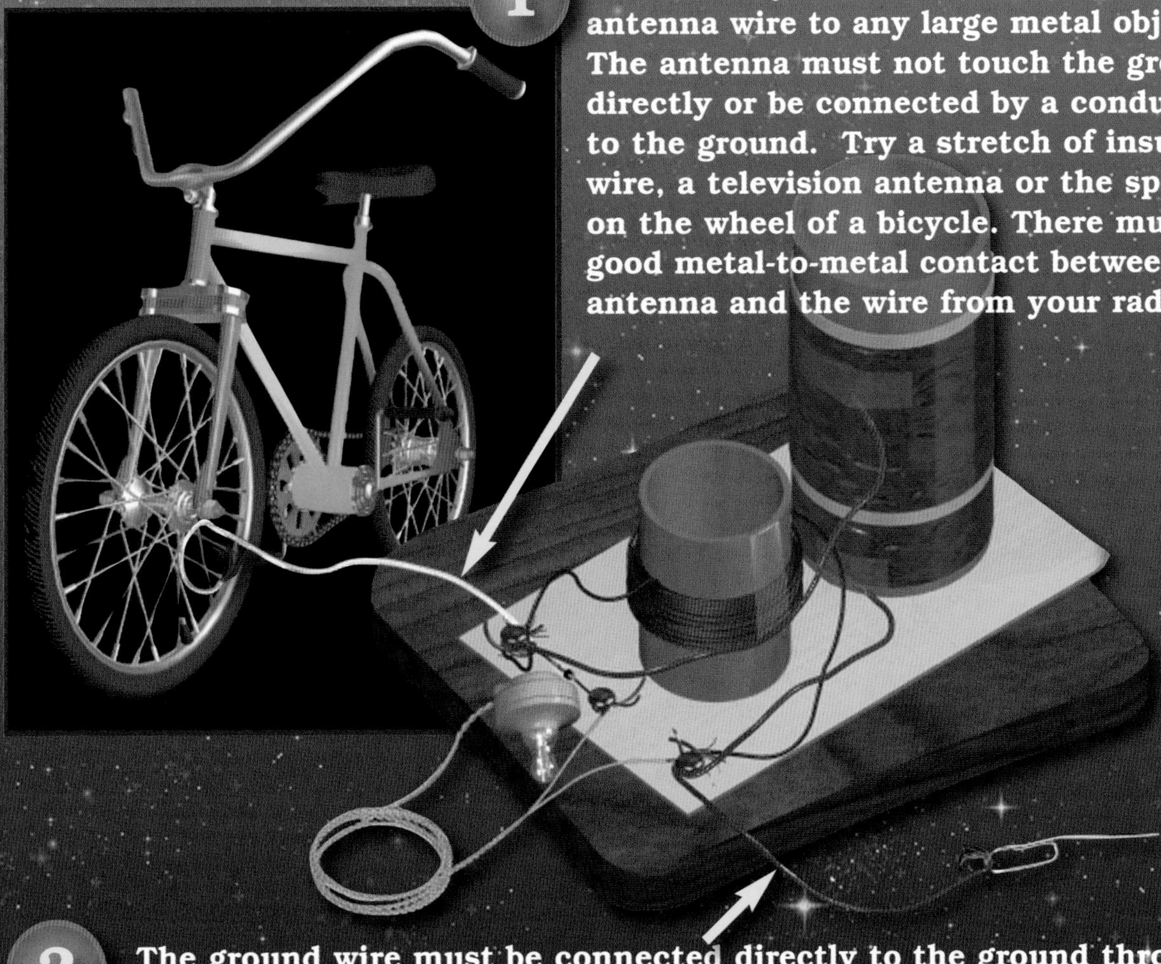

2. The ground wire must be connected directly to the ground through a conductor. An EXCELLENT ground is the metal case found on the back of a computer which is plugged into the wall. A metal plumbing pipe or heating duct may work too.

3. Once your radio is connected to both an antenna and to ground, put the earphone in your ear.

Slide the aluminum foil capacitor up and down slowly until you find a station.

Listen carefully. *What do you hear?*

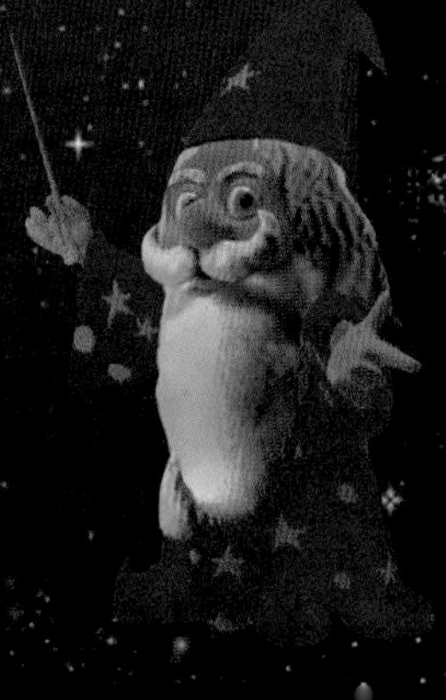

HOW THE RADIO WORKS

You live in a sea of radio waves. The colored lines represent radio waves.

The GREEN circles represent signals coming from one radio station's transmitter. The BLUE, YELLOW and RED waves represent transmissions from other stations.

Real radio transmitters don't send out color coded signals, they send out frequency coded signals.*

The task of your radio is to pick out ONE radio frequency and turn it into sound.

The antenna turns radio waves into electrical current. The antenna picks up all the radio waves, it's not picky.

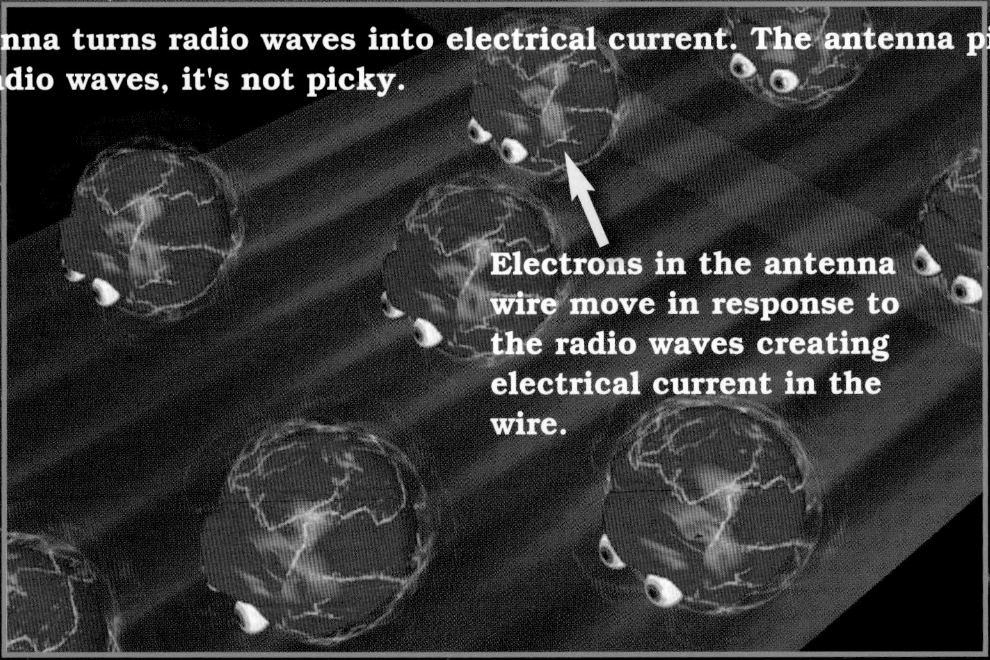
Electrons in the antenna wire move in response to the radio waves creating electrical current in the wire.

The ground wire allows current to flow through the circuit of the radio.

* Visit www.sciencewiz.com for a more complete explanation about radio waves.

The aluminum foil and paper on the tube form a capacitor. A capacitor is any two conductors separated by an insulator. In the capacitor you made, the conductors were two pieces of aluminum foil and the insulator was a piece of paper.

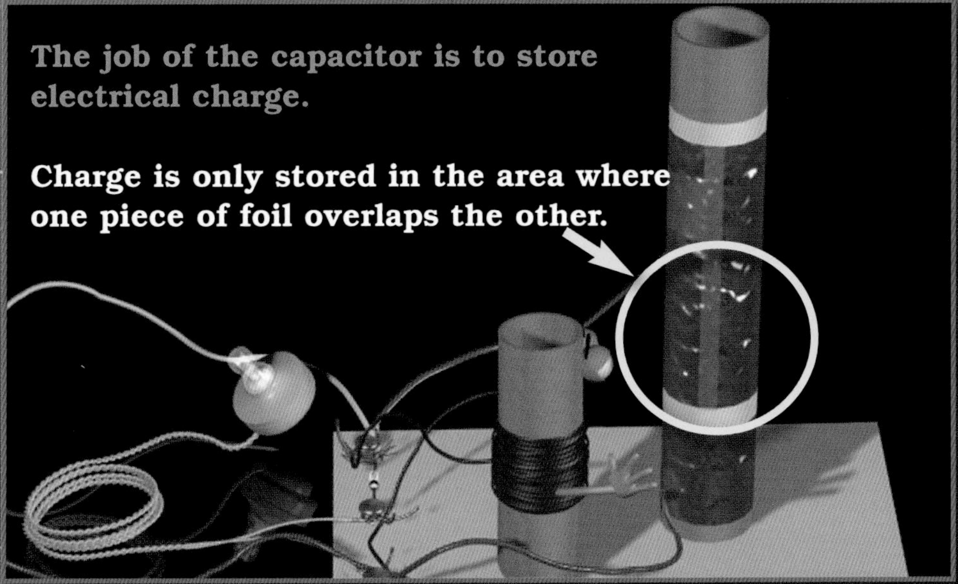

The job of the capacitor is to store electrical charge.

Charge is only stored in the area where one piece of foil overlaps the other.

The electrons collect first on the outside piece of aluminum foil and then flow away and collect on the inside piece of aluminum foil. The electrons flow back and forth, back and forth.

Your homemade capacitor is a variable capacitor.

As you slide the outer foil down, the amount of foil facing the inner layer INCREASES and the amount of charge that the capacitor can store also INCREASES.

As you slide the outer foil up, you lessen the amount of foil facing the inner layer and you LOWER the amount of charge that can be stored.

The capacitor has a teammate, the COIL. The coil in the radio is no longer just an electromagnet, it has a whole new job.
It is an INDUCTOR.

The inductor slows down the change in the current. Current can change speed. Cars can change speed. What does a change in speed mean? The idea of a change in speed, speeding up or slowing down, is important to the understanding of how an inductor works.

SPEEDING UP

SLOWING DOWN

The mission of the inductor is to slow down the change in the current.

In a straight wire the current can change its speed very quickly.

In a COIL the current takes longer to speed up or slow down. The more turns the coil has, the longer it takes for the current to change.

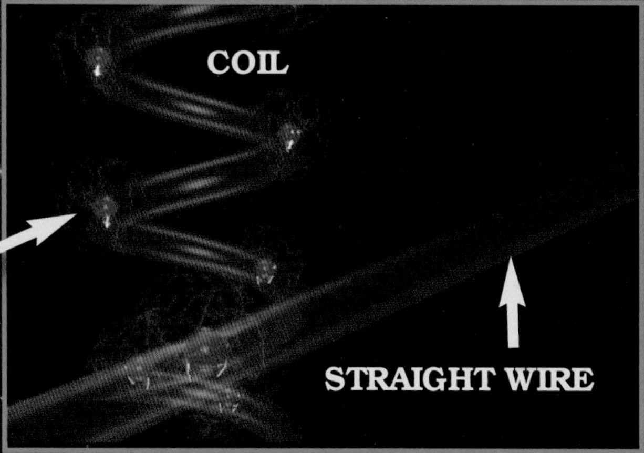
COIL
STRAIGHT WIRE

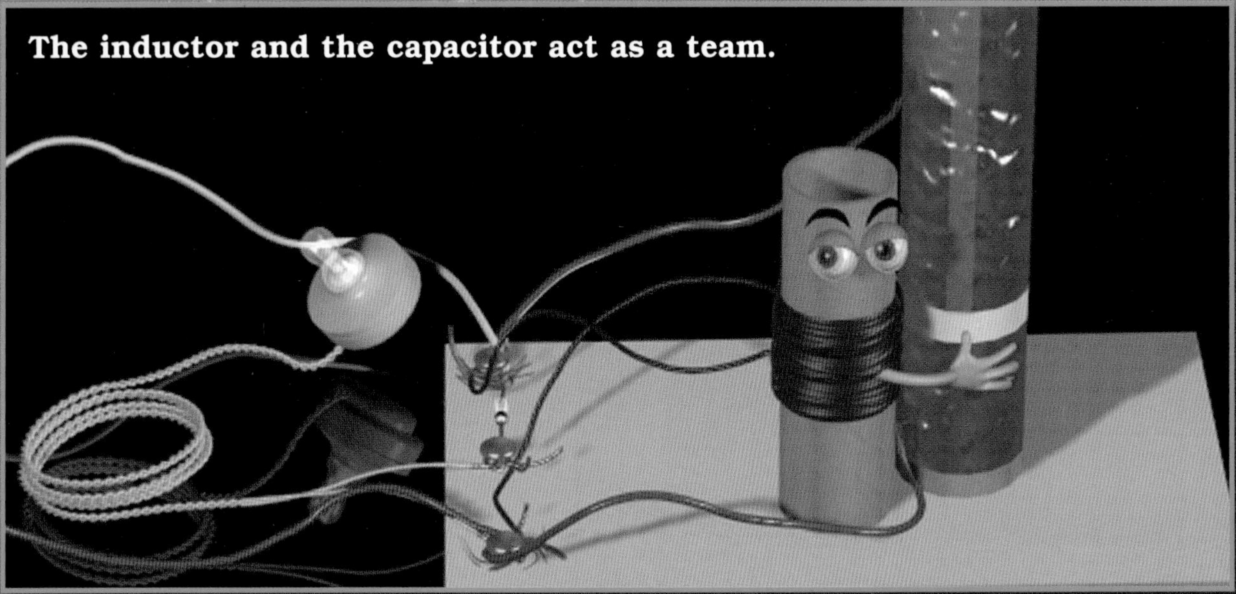
The inductor and the capacitor act as a team.

Remember the antenna picks up all the radio waves it can find, it is not picky. The coil and capacitor team is picky.

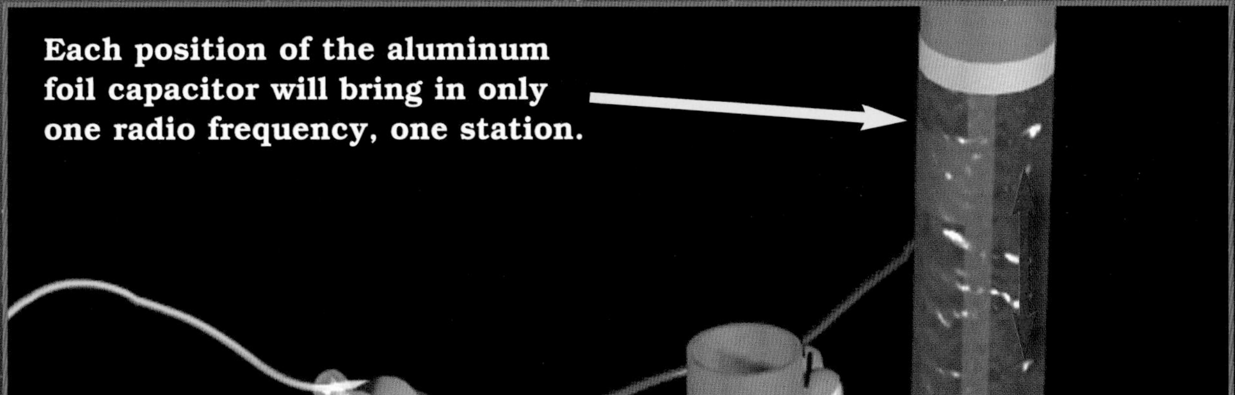
Each position of the aluminum
foil capacitor will bring in only
one radio frequency, one station.

The team tunes to one station, it resonates with that station's frequency and no other.

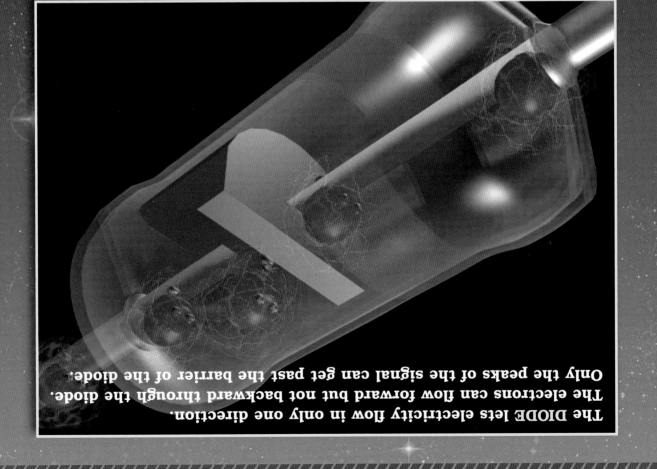

The **DIODE** lets electricity flow in only one direction.
The electrons can flow forward but not backward through the diode.
Only the peaks of the signal can get past the barrier of the diode.

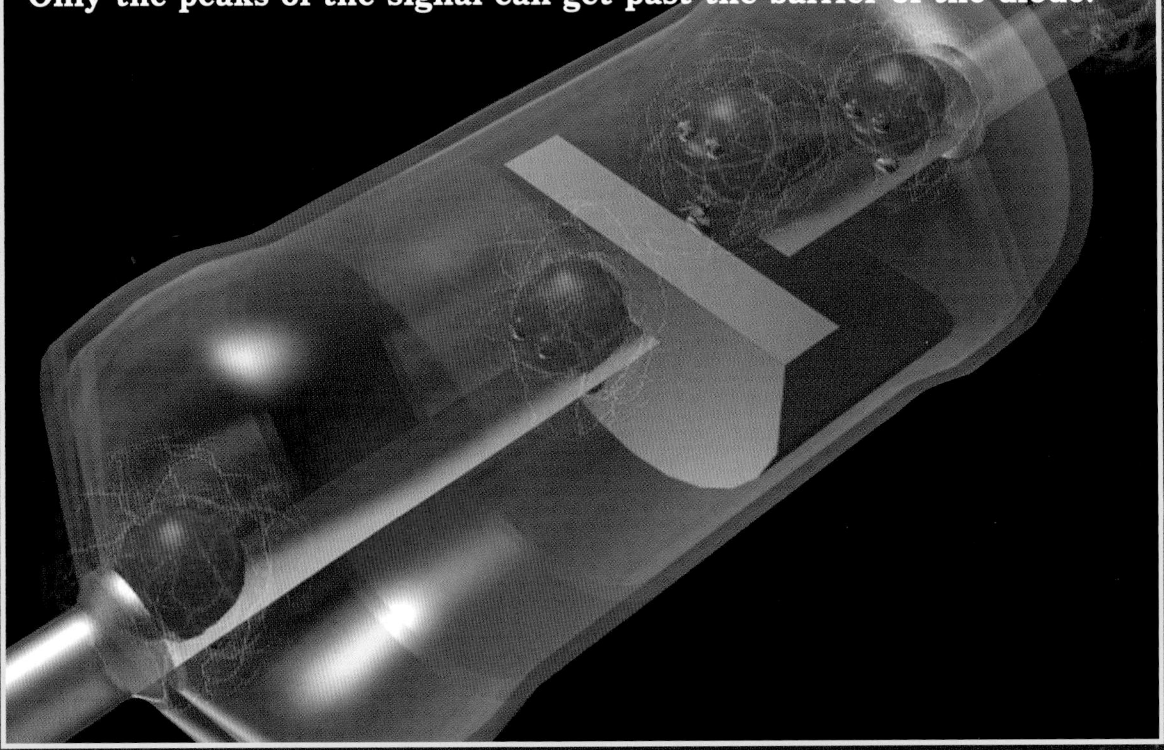

The earphone blends together the peaks of the signal coming out of the diode and turns them into **SOUND VIBRATIONS**.

What in the earphone creates sound vibrations?

Let's look inside the earphone.

1 There is a coil and a magnet in the earphone.

COIL
MAGNET

2 As electricity flows through the coil, it creates an electromagnet.

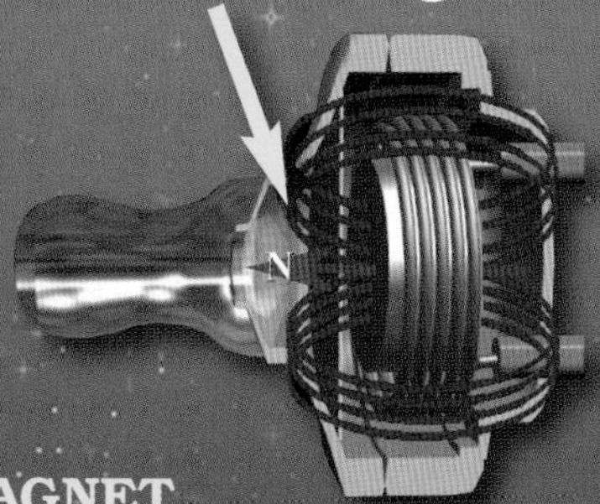
N

3 In the earphone is a diaphragm with iron particles imbedded in it.

As the current goes up and down through the coil, the diaphragm is attracted more or less to the electromagnet.

The diaphragm vibrates creating....

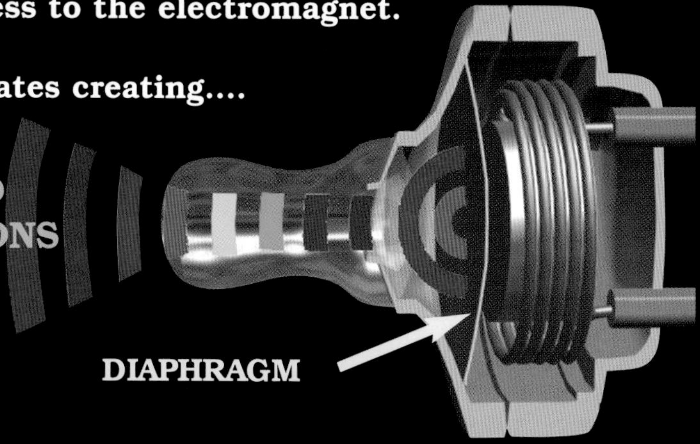
SO-UND
VI-BRA-TIONS
DIAPHRAGM

Now let's put it all together!

Radio stations transmit invisible electromagnetic waves called radio waves. The antenna picks up all the radio waves and turns them into electrical current. The capacitor-inductor team picks out one signal, one station, and ignores all the other signals. The diode clips the peaks of the signal. The speaker puts the peaks together and turns them into sound vibrations. From your ear to your brain, the vibrations become sound.

BONUS WEB MATERIALS

Go to www.ScienceWiz.com to build an FM Radio.

Go to www.ScienceWiz.com to earn your AWARDS.

AWARD WINNERS

THIS BOOK IS ONE OF A SET OF AWARD WINNING TITLES, AWARDS FOR THIS BOOK INCLUDE SCIENTIFIC AMERICAN YOUNG READERS BOOK AWARD AND NATIONAL PARENTING MAGAZINE AWARD, THE GOLD

Acknowledgements

The project layouts were created for the CD-ROM ElectroWizard™ Inventions, the design of which was funded by the National Science Foundation.

Computer generated 3-D artwork: Art Huff

Editing: Ann Einstein and Rachael Norman

Special thanks to the Exploratorium in San Francisco for allowing us to test materials on-site. Thank you to the teachers and parents of Sheldon Elementary School, West Contra Costa Unified District, Richmond, California, who were so dedicated in their support of the testing of these materials. Without Karen Moorhead's, Charlotte Lininger's, Judith Coleman's, and Buddy Philips' contributions to this project over a number of years, these materials would not reflect the perspective of the elementary age child. Thank you to all the child testers.

Additional Copies

Additional copies of this book and replacement parts, as well as a free catalogue of our other books and materials, are available. Comments are welcomed.

Published by Norman and Globus, Inc.
P.O. Box 20533
El Sobrante, CA 94820-0533
(510) 222-2638 FAX (510) 223-6953
website: www.sciencewiz.com e-mail: drpenny@sciencewiz.com

The book and carton are printed in China.
Components are manufactured in the U.S.A. and China.

ISBN 978-188697803-4
UPC 630227078034 **09 / 2018**